Practical Manual Series - 2

Postharvest Technology of Horticultural Crops

Practical Manual Series - 2

Postharvest Technology of Horticutural Crops

S.K. SHARMA
Assistant Professor, Horticulture
(Fruit & Vegetable Processing)
College of Forestry & Hill Agriculture
G.B. Pant University of Agriculture & Technology
Hill Campus, Ranichauri, Tehri (Garhwal) Uttarakhand-249199

M.C. NAUTIYAL
Dean
College of Forestry & Hill Agriculture
G.B. Pant University of Agriculture & Technology
Hill Campus, Ranichauri, Tehri (Garhwal) Uttarakhand-249199

New India Publishing Agency
Pitam Pura, New Delhi- 110 088

Published by
Sumit Pal Jain *for*

New India Publishing Agency

101, Vikas Surya Plaza, CU Block, L.S.C. Mkt.,
Pitam Pura, New Delhi- 110 088, (India)
Phone: 011-27341717, Fax: 011-27341616
E-mail: newindiapublishingagency@gmail.com
Web: www.bookfactoryindia.com

ISBN : 978-81-90851-20-6

Composed and Designed by NIPA

डॉ. एच.पी. सिंह
उप महानिदेशक
(बागवानी)

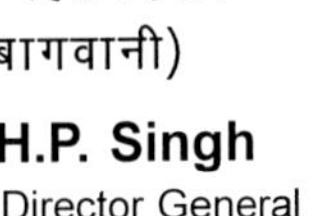

Dr. H.P. Singh
Deputy Director General
(Horticulture)

भारतीय कृषि अनुसंधान परिषद
कृषि अनुसंधान भवन–II, पूसा, नई दिल्ली–110 012

INDIAN COUNCIL OF AGRICULTURAL RESEARCH
Krishi Anushandhan Bhavan-II
Pusa, New Delhi-110 012

FOREWORD

Horticulture comprising of fruits, nuts, vegetables, tubers, mushrooms, spices, floriculture, medicinal and aromatic plants and plantation crops, has emerged as one of the potential enterprises in accelerating the economic growth. It offers not only wide range of options to farmers for diversification but also provides ample scope for sustaining large number of agro- industries, which generate employment opportunities, add value to produce and thereby support livelihood to many. The health conscious population is demanding more fruits, vegetables and their value added products, thus, there is an increasing demand for horticultural produce.

Although there has been many fold increase in production, but post harvest losses continue to be high. Therefore, effective post harvest management assumes a significance. Many technologies have been developed for grading, packing, storage and value addition which require to be put at a place to make it handy for the uses. The publication entitled "Post-harvest Technology of Horticultural Crops - A Practical Manual" covers a great deal of information on practical aspects for reducing post-harvest losses and adding value through processing.

I am sure the book shall be of immense value to all those who are concerned with post-harvest handling, marketing, and processing of fruits and vegetables. Finally I complement the authors for their sincere efforts.

(H.P. Singh)
DDG (Horticulture)
ICAR, New Delhi

PREFACE

Postharvest Technology is a very important branch of agriculture and its importance further increases in the field of horticulture, as the horticultural crops are highly perishable in nature. This necessitates undertaking continuous research on the preservation of these perishables. As such, there is no book or manual available which discusses the basics of undertaking laboratory or research experiments on various topics pertaining to the postharvest management of horticultural crops. The students, teachers and researchers often search a direct reference which is complete on the subject for an undergraduate or postgraduate student.

The authors, therefore, made an attempt to meet out this need. This practical manual on "Postharvest Technology of Horticultural Crops" has been divided into 33 exercises covering various aspects of fresh fruit handling, acquaintance with various equipments and machines, preparation of value added products, nutritional analysis of various products, sensory quality analysis of various products etc. Effort has been made to put the contents in such a way that even any teacher who has only a superficial knowledge on the subject can also teach or conduct practical class after going through the information given in each exercise. The information on the FPO, significance of Triangle methods, conduct of a student or researcher in the laboratory maturity indices, conversion tables, limits of food additives etc. have also been given under the Annexures in the manual.

We hope that the students, researchers, teachers and all those who have interest in the subject would welcome the manual and find it very useful. We thankfully acknowledge the help obtained from various books (mentioned in the reference section) that were consulted during the preparation of different exercises given in this manual. Although, efforts were made to present the information in most understandable manner, in simple language and with minimum of errors and omissions, valuable suggestions from the users of the manual, especially students, researchers and teachers will be appreciated to improve it further.

AUTHORS

CONTENTS

Exercise 1

Acquaintance with Different Equipments and Machinery Used in Postharvest Technology

LEARNING OBJECTIVES

The students should be able to

- ❒ Identify different equipments and machinery used in Postharvest Technology laboratory.
- ❒ Understand the principles and methods to use the equipment or machinery.

Materials Required

Equipments and machinery used in Postharvest Technology laboratory for processing, packaging and quality analysis.

Procedure

1. Identify the equipments and machinery, note down the typical features about its shape, size, colour etc.

2. Note down the principle of their functioning
3. Learn how to use the equipments and machinery

Observations

S. No.	Name of the equipment or machinery	Typical features about its shape, size, colour etc	Principle of functioning	Use
1				
2				
3				
4				
5				
6				
7				
8				
9				
10				
11				
12				
13				
14				
15				

For Teacher

a. The teacher shall take the students to the laboratory and make them identify different equipments and machinery and tell them the principles and use of the items identified.

RELEVANT INFORMATION

The photographs and uses of some of the common equipments and machinery used in a Postharvest Technology Lab are given ahead.

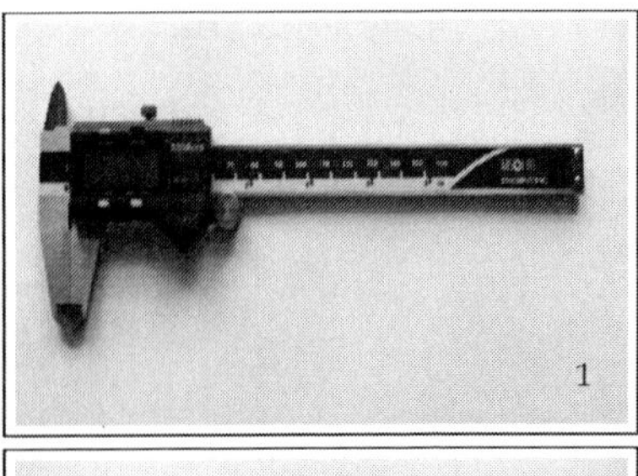
1

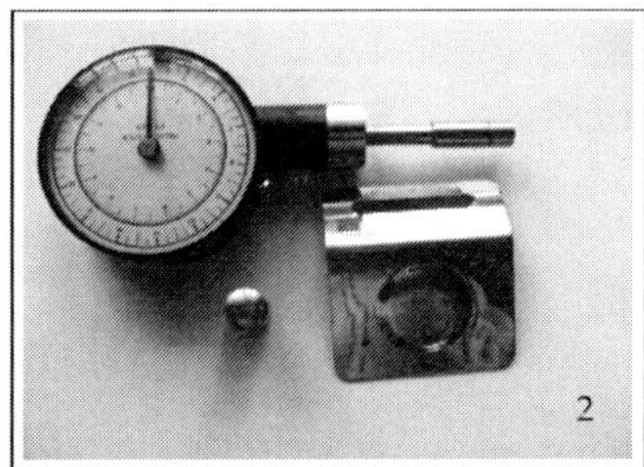
2

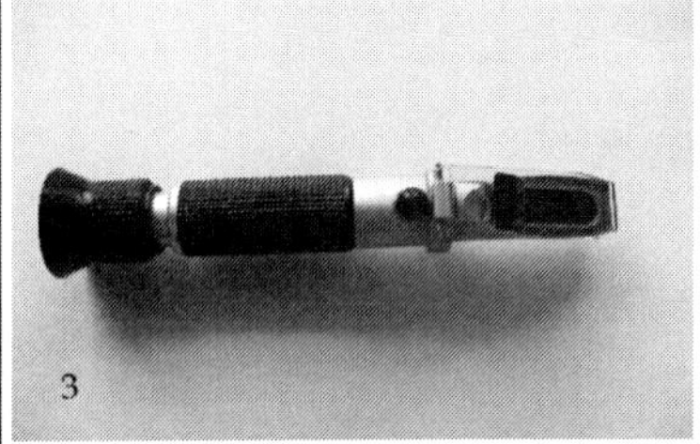
3

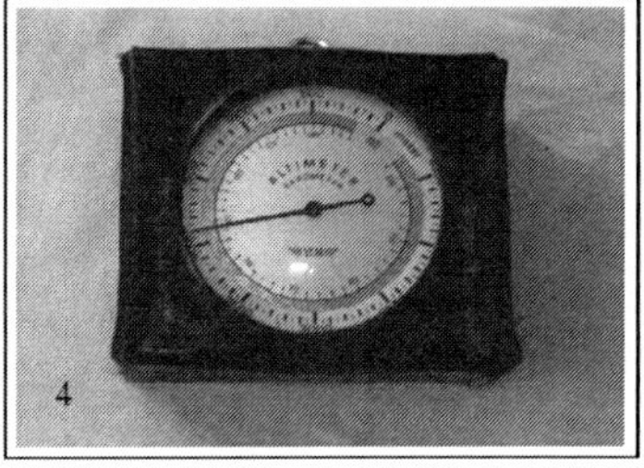
4

5

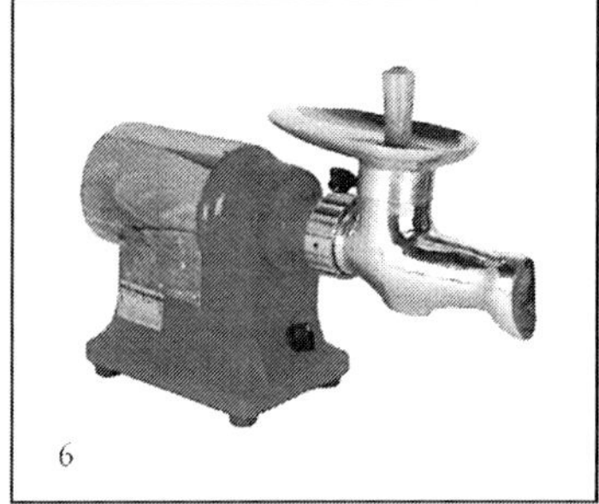
6

1. **Vernier Calliper:** Used to measure the thickness or diameter of fruits or given samples.
2. **Penetrometer:** Used to measure the firmness / pressure of fruits. It is also called as fruit pressure tester. The firmness is measured and expressed in lbs/in^2 or Kg/cm^2.
3. **Hand Refractometer:** It is used to measure the TSS of juices, fruits, syrups etc. The TSS (total soluble solids) are expressed in ^{o}Brix or per cent. TSS is measured at room temperature and corrected using correction Tables.
4. **Altimeter:** It is used to measure the altitude of any place above mean sea level. The measurement is generally expressed in terms of m amsl. It works on the principle that the atmospheric pressures reduce as we go up above mean sea level.
5. **Screw Type Juice extractor (Hand Operated):** Used for extraction of juice from citrus fruits. Runs when force is applied manually. The juice filtered through a sieve is obtained separately and seeds and residue are obtained separately.
6. **Screw Type Juice extractor (Motor Operated):** Same as item at No. 5 but it runs with the help of electric power supply.

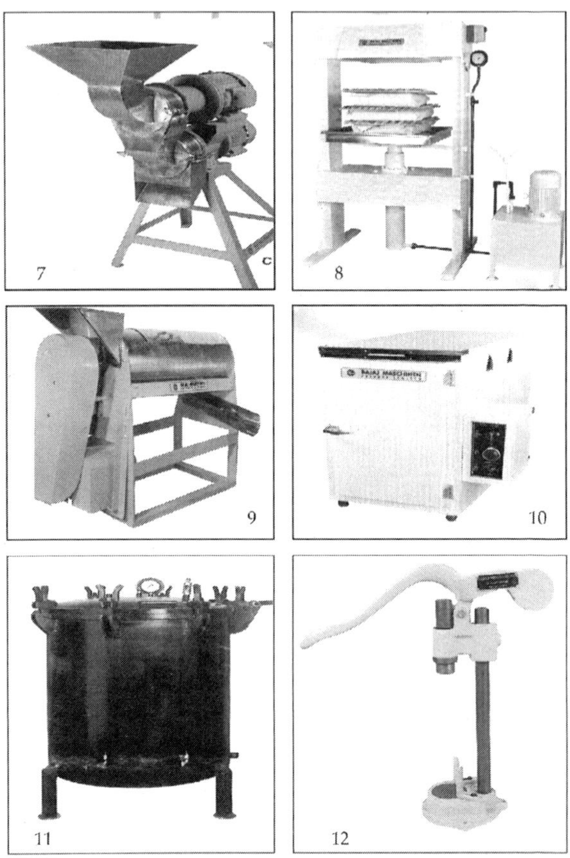

Courtesy: M/s Bajaj Maschinen Pvt. Ltd., Ghaziabad, U.P.

7. **Fruit Mill:** Used for grating of fruits i.e. apple, pear etc.
8. **Hydraulic Press:** Used for extraction of juice from fruits i.e. apple, pear etc. Fruit pieces are first grated in a Fruit Mill or Grater and then put in cloth in alternate layers. The hydraulic pressure is maintained either manually or through motor driven by electric power.
9. **Pulper:** Used for extraction of pulp from fruits. Fruits after making pieces and addition of small quantity of water are generally heated for about 10-20 min. so as to make the pieces soft. The boiled material is passed through the pulper to separate the pulp and seeds & skin.
10. **Dehydrator:** Used for preparation of dried products. Drying in dehydrator is generally done at temperatures of 55-60° C. Higher temperatures may lead to burning of the material.
11. **Autoclave:** Used for sterilization of samples at increased pressures. 10 lbs pressure = 115°C, 15 lbs pressure = 121.1° C.
12. **Crown corking machine:** Used for corking the bottles and sealing them air tight.

Courtesy: M/s Bajaj Maschinen Pvt. Ltd., Ghaziabad, U.P.

13. **Can Body Reformer:** Used for reforming the body of cans into cylindrical shape
14. **Flanger:** Used for making flanges on the cylindrical Can body so as to hold the lids
15. **Can Body beader:** Used for making beads of different shapes on the Can body.
16. **Double Seamer:** Used for hermatic sealing of cans from both sides. This has two roller operations. In the first roller operation the lid is clinched into the can body but air can still pass through. The second roller operation of the double seamer presses the lock in such a manner that it becomes air tight.
17. **Exhaust Box:** The partially sealed cans after the first roller operation of the double seamer are passed through the exhaust box in which high temperature are maintained by passing steam. During this process all the dissolved air from the contents of the filled can is removed. The cans after exhausting are sealed immediately in the Double seamer using the second roller operation.
18. **Bottle washing machine:** Used for washing glass bottles
19. Automatic bottle filling machine: Used for filling the prepared liquid or semi solid product in bottles
20. **Lug cap sealing machine:** Used for sealing jars having lug caps i.e. jam jars.

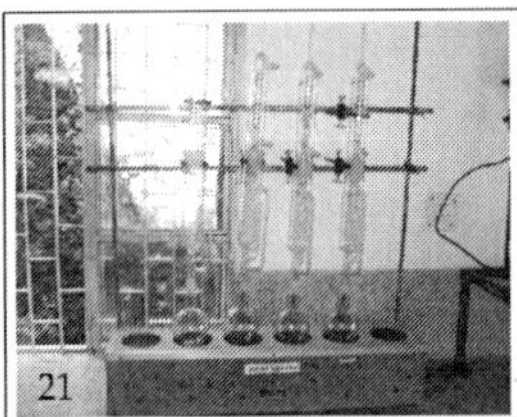

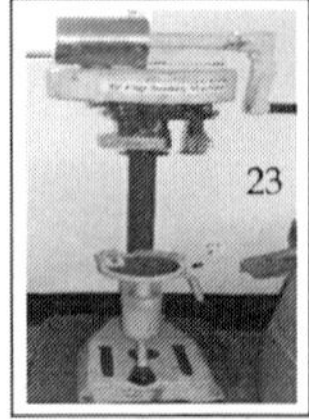

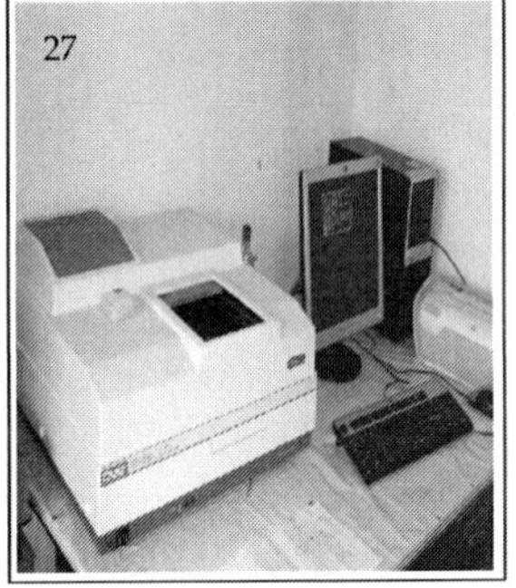

21. **Soxhlet Apparatus:** It is used to estimate fat content of the powdered samples
22. **Hot Water Bath:** It is used to conduct experiments at fixed temperature. The temperature can be regulated by the thermostat.
23. **PP Cap Sealing Machine:** Pilferage Proof Cap sealing machine. It is used to seal glass bottles eg. squash bottles with PP screw caps.
24. **Oil Expeller:** Used for extraction of oil from fruit kernels, etc. on commercial scales
25. **Filter Press:** Used for the clarification of extracted oils or juices during enzymatic clarification.
26. **Decorticator:** It is used for breaking the hard seeds of stone fruits i.e. apricot, plum, peach etc.
27. **UV Vis spectrophotometer:** Used for estimation of various nutrient elements in fruit products and for quality analysis of fruits and fruit products.
28. **Low cost Poly tunnel Drier:** Used for dehydration of various fruit products using solar heat.

Exercise 2

To Study the Maturity Indices in Various Fruits and Vegetables

Learning Objectives

The students should be able to

- ❒ Understand classification of maturity indices
- ❒ Learn various maturity indices used in different fruits and vegetables.

Materials Required

Fruits/vegetables, hand refractometers, fruit pressure tester, weighing balance, scale, chemicals and glassware required for analysis.

Procedure

1. Go to the orchard or vegetable field and collect samples of fruits and vegetables
2. Apply the specific maturity indices to the samples

3. Determine whether the samples are mature, immature, over mature etc.
4. Harvest the fruits
5. Count No. of unmarketable fruits in every 100 fruits harvested

Observations

S No.	Crop	Maturity index used	Reading/ observation	Maturity class	Unmarketable fruits (%)
1					
2					
3					
4					
5					
6					
7					
8					
9					
10					
11					
12					
13					
14					
15					

For Teacher

a. The teacher shall take tell the students the classification of maturity indices given as follows
b. The teacher shall also teach the students various definitions of maturity

c. The teacher shall take the students for collection of fruits or vegetable samples at different stages of maturity

d. He / she shall ask the students to apply specific maturity index and record the observations

RELEVANT INFORMATION

Classification of Maturity Indices

1. Computational Methods	:	Calendar date (say 5th June or 26th July)
	:	Days from full bloom to harvest (say 141 days)
	:	Mean heat units
2. Physical Methods	:	Fruit Retention Strength (Kg)
	:	Fruit Size, Shape, Colour etc.
	:	Weight (g)
	:	Specific Gravity
	:	Colour (Skin, pulp and seed)
	:	Flesh Firmness (Kg/cm^2)
	:	TSS (^{o} Brix or %)
	:	Juice Content (%)
	:	Oil Content (%)
	:	Tenderness (peas, beans)
	:	Abscission layer
	:	Surface gloss (plum)
3. Chemical Methods	:	Titratable acidity (%)
	:	TSS-Acid Ratio

	:	Sugars (%)
	:	Sugar-Acid Ratio
	:	Starch (%)
4. Physiological Methods	:	Respiration Rate ($mlCO_2Kg^{-1}h^{-1}$)
	:	Ethylene Evolution Rate (μL/L)

Maturity Definitions

Physiological maturity: Stage at which a plant or a plant part continue ontogeny even if detached.

Ontogeny: Complete developmental history of an organism from egg/spore/bud etc. to an adult individual.

Horticultural maturity: Stage at which a plant or a plant part possesses all the pre-requisites for utilization by the consumer for a particular purpose.

Suggested Readings

Sharma RM and Singh RR 2000. Harvesting, postharvest handling and physiology of fruits and vegetables. *In:* Postharvest Technology of Fruits and Vegetables – Handling, Processing, Fermentation and Waste Management. Verma LR, Joshi VK (eds.) Indus Pub. Co. New Delhi, pp 94-147.

Wills RBH, McGlasson WB, Graham D, Lee TH and Hall EG 1996. Postharvest – An Introduction to the Physiology and Handling of Fruit and Vegetables. CBS Pub. Distributors, New Delhi p174.

□□□

Exercise 3

To Study the Effect of Pre-Cooling on Storage Life of Fruits and Vegetables

LEARNING OBJECTIVES

The students should be able to

- Understand the concept of pre-cooling in fruits and vegetables
- Study whether pre-cooling does have any impact on shelf life or not ?
- Layout an experiment for the study of the effect of pre-cooling on shelf life and quality

Materials Required

Fruits/vegetables, thermometer, refractometers, NaOH, Phenolphthalein, glassware, weighing balance, plastic crates for storage of commodity, water cooling facility.

Procedure

1. Harvest the fruit / vegetable
2. Count No. of unmarketable fruits in every 100 fruits harvested
3. Record the internal fruit temperature in about 5-10 randomly selected fruits with the help of a thermometer
4. Pre-cool the commodity in cold water (<5°C), air or refrigerator till the temperature reaches near its optimum storage temperature
5. Record the internal fruit temperature
6. Record the time required to pre-cool the commodity
7. Keep the product for storage at ambient conditions alongwith control
8. Record the difference in the shelf life and quality during storage

Observations

S. No.	Parameter	Control fruits/ vegetables	Pre-cooled fruits/ vegetables
1	Internal fruit temp. at harvest		
2	Internal fruit temp. after pre- cooling		
3	Time taken for pre-cooling (hrs)		
4	Initial weight at beginning of storage (Kg)		
5	Weight loss (%) after 2, 4, 6, 8 ....days		
6	Rotting (%) after 2, 4, 6, 8 ....days		
7	TSS (° Brix) at 0, 2, 4, 6, 8 ...... days		
8	Acidity (%) at 0, 2, 4, 6, 8 ....days		

Observations at S No. 5,6,7,8 shall depend upon the type of fruit and its expected storage duration. The time interval can be adjusted accordingly.

For Teacher

a. The teacher shall tell the students about the definition and methods of pre-cooling.

b. He/she shall take the students to the field for fruit harvesting

c. He/she shall tell them about the expected shelf life of the harvested fruit / vegetable and the time duration at which the observations are to be recorded.

c. He/she shall ask the students to record observations as detailed above.

RELEVANT INFORMATION

Pre-cooling: Prompt cooling of the commodity immediately after harvest (generally within 24 hrs) which aims at removal of field heat.

Cooling Methods

Room cooling	All fruits and vegetables (Banana, beans, cabbage, coconut, garlic, ginger, lime, lemon, onion, orange, cucumber, pineapple, potato, pumpkin, radish, sweet potato, tomato, watermelon etc.)
Forced air cooling	Fruits and fruit type vegetables (banana, Barbados cherry, berries, brussels sprouts, cucumber, eggplant, fig, ginger, grape, guava, litchi, mango, kiwifruit, mushroom, okra, oranges, papaya, passion fruit, bell pepper,

contd...

Table contd...

	pineapple, persimmon, pomegranate, quince, sapota, strawberry, tomato, tubers, cauliflower etc.)
Hydro cooling	Stem and leafy vegetables, fruits (artichoke, asparagus, beet, broccoli, brussels sprouts, cantaloupe, carrot, cauliflower, celery, Chinese cabbage, cucumber, brinjal, green onion, kiwifruit, leek, knol-khol, orange, parsley, parsnip, peas, pomegranate, radish, spinach, rhubarb, swiss chard, summer squash, water cress etc.)
Ice cooling (Package icing)	Roots, stems, flower type vegetables (endive, broccoli, Brussels sprouts, carrot, spring onions, Chinese cabbage, leek, parsley, snow peas, spinach, swiss chard, sweet corn etc.)
Vacuum cooling	Some stem, leafy and flower type vegetables (endive, brussels sprouts, carrot, cauliflower, Chinese cabbage, celery, leek, mushrooms, lima bean, spinach, sweet corn, swiss chard etc.)
Transit icing	All fruits and vegetables

Note: - *Some fungicides may be mixed in water during hydro-cooling to reduce decay incidence*

- *Weight loss during forced air cooling can be reduced by maintaining high (95%) relative humidity in the pre-cooling chamber.*

Suggested Readings

Mitchell FG 1985. Cooling Horticultural Commodities *In:* Postharvest Technology of Horticultural Crops. Kader AA, Robert RF, Mitchell FG, Reid MS, Sommer NF, Thompson JF (eds.) Coop. Ext.Univ. Calif. DANR Berkeley 94720 pp35.

Sharma RM and Singh RR 2000. Harvesting, postharvest handling and physiology of fruits and vegetables. *In:* Postharvest Technology of Fruits and Vegetables - Handling, Processing, Fermentation and Waste Management. Verma LR, Joshi VK (eds.) Indus Pub. Co. New Delhi, pp 94-147.

Srivastava RP and Kumar S 2002. Fruit and Vegetable Preservation - Principles and Practices. International Book Distributing Co., Lucknow. p 474.

General recommended storage conditions for fruits and vegetables

Fruit/vegetable	**Temperature (°C)**	**Relative humidity (%)**	**Approx. storage life**
Asparagus	0	95	3 – 4 weeks
Beans green	4 – 7.2	90 – 95	1 – 2 weeks
Bitter gourd	0.6 – 1.7	85 – 90	1 month
Brinjal	8 – 12	90 – 95	1 – 3 weeks
Broccoli	0	95	1.5 – 2 weeks
Cabbage	0 – 1.7	92 – 98	3 – 6 weeks
Carrot	0	95	20 – 24 weeks
Cauliflower	0 – 1.7	85-98	3 – 7 weeks
Chilly	7	90-95	2-3 weeks
Colocasia	11.1 – 12.8	85-90	21 weeks
Cucumber	10 - 13	92-95	1.5 – 2 weeks
Garlic	0	65 - 70	6 – 9 months
Ginger	7.2 - 13	65 -75	4 – 6 months
Lettuce leaves	0	95-98	1 - 3 weeks
Mushroom	0	95	3 – 4 days
Muskmelon	1.7 – 7.2	85 -90	1.5 – 4.5 weeks
Okra	8 - 9	90	2 weeks
Onion	0	70 - 75	5 -6 months
Pea	0	88 - 98	1-3 weeks
Potato	3 - 5	85	5 – 10 months
Pumpkin	1.7 – 11.6	70 - 75	6 – 9 months

contd...

Table contd...

Radish	0	88 – 92	3 – 5 weeks
Spinach	0	95 -100	10 -14 days
Tomato	7.2 - 10	85 - 90	1 - 4.5 weeks
Turnip	0	90 - 95	8 – 16 weeks
Watermelon	7.2 – 15.6	80 -90	2 weeks
Apple	0 – 4	80 – 90	4 – 8 months
Apricot	0 – 2	80 – 85	1 – 2 weeks
Banana	12 – 13	80 – 90	1 – 2 weeks
Cherry	0 – 2	85-90	2 weeks
Dates	7 - 8	85- 90	2 weeks
Fig	0 - 2	85 - 90	4 weeks
Grape	-1.1 - 2	80-95	3-8 weeks
Guava	7.2 - 10	85- 90	2 – 3 weeks
Jackfruit	11 - 13	85 - 90	1.5 months
Kiwifruit	-0.5 - 0	90 - 95	3 – 5 months
Lemon	8 - 10	85 -90	1 - 6 months
Lime	8-10	85 -90	3 - 6 weeks
Litchi	0 – 2.1	85-95	3-10 weeks
Mandarin	5 - 8	85 - 90	10 -14 weeks
Mango	8 - 12.8	85 -90	2 – 7 weeks
Papaya	7.2 – 10	80-90	1-2 weeks
Passion fruit	7 - 8	80 - 85	4 – 5 weeks
Pear	0 – 1	85 -90	3 – 6 months
Peach	0 – 3	85 - 90	2 – 4 weeks
Persimon	0 - 2	85 - 90	7 weeks
Pineapple	8 - 13	85 - 90	3 – 6 weeks
Plum	0 – 2	85 - 90	4 – 8 weeks
Pomegranate	0 – 2	80 - 90	2 – 6 weeks
Sapota (Chiku)	3 – 4	85 - 90	6 – 8
Strawberry	0 – 2	80 -95	5-7 days

❒❒❒

Exercise 4

To Study Various Pre-Treatments Given to Fruits and Vegetables before Storage and Marketing

LEARNING OBJECTIVES

The students should be able to

- Learn various pre-treatments given to fruits and vegetables before storage
- Understand the purpose of giving a particular treatment and the principle by which the treatment cause its effect

Materials Required

Fruits/ vegetables, washing tanks/ containers, size grader, fruit waxes, packaging material, neem based fungicides etc.

Procedure

1. Take available fruits or vegetables
2. Wash them in plain water or chemical solution as told by the teacher

3. Shade dry the fruits / vegetables
4. Prepare suitable wax solution
5. Apply wax on to the fruits as told
6. Record the fruit weight before packing
7. Pack the fruits properly in the trays / boxes depending upon the type and purpose
8. Do curing if required
9. Do pre-packaging, if desired
10. Label the package for desired information
11. Keep the fruits for storage
12. Record the observations pertaining to weight loss and rotting periodically depending upon the commodity and its shelf life.

Observations

A. Name of Fruit

B. Information regarding pre-treatments

Pre-treatments	Given or not	Details of treatment
a. Washing		Give the details of the solution prepared for washing
b. Trimming		Record how much portion (%) of the commodity was trimmed off
c. Sorting		For what characteristics the commodity was sorted?
d. Grading		Indicate the % of fruits obtained in a particular grade out of the whole lot or calculate No. of fruits of a particular grade in a lot of 500 or 1000.

e. Chemical treatment	Record the name of chemical used for postharvest dip, concentration used and time of dipping
e. Shade drying	What was the room temperature and how much time did it take for drying ?
f. Curing	What was the temperature during curing and what changes were observed after curing ?
g. Pre-packaging	What was the size / weight of pre-package and what was the packaging material used ?
h. Final packaging	What material was used for packing, No. of trays required per box, No. of fruits accommodated per tray, No. of fruits accommodated per box, weight of box after packing ?
i. Labeling	What was the information put on the label of the package ?

C. Fruit behaviour during storage

Parameter	**Days or months in storage**						
	Initial	**2**	**4**	**6**	....	....	......
Total No. of fruits							
Weight (Kg)							
Weight loss (%)							
No. of Rotted fruits							
Rotting (%)							
Shrivelling							
TSS (°Brix)							

For Teacher

a. The teacher shall tell the students about various pre-treatments given to fruits and vegetables

b. He/ she shall also explain in detail the methods of applying the pre-treatments, and why they are applied.

c. He / she shall ask the students to record various observations pertaining to the given pre-treatments on the given format.

RELEVANT INFORMATION

Washing : It is done in plain water or chlorinated water or fungicides i.e. diphenyl amine (0.1-0.25%) & Ethoxyquin (0.2-0.5%) as postharvest dip

Sorting and Grading : This may be done manually or mechanically.

Size grades of apple

Grade	Min. Dia (mm)
Super Large	85
Extra Large	80
Large	75
Medium	70
Small	65
Extra Small	60
Pitoo	55

Size grades of apricot

Grade	Min. Dia (mm)
Special	42
Grade I	36-42
Grade II	< 36

Size grades for Malta oranges

Grade	Min dia. (inches)	Min. dia. (cm)
Extra special	3.25	8.255
Special	3.00	7.620
Good	2.75	6.985
A	2.50	6.350

Size grades for pomegranate

Grades	Fruit weight (g)	Fruit Characteristics
Super	> 750	Attractive, very large, dark red in coloured, without blemishes
King	500-750	Attractive, large without blem ishes
Queen	400-500	Large Medium, attractive without blemishes
Prince	300-400	Medium attractive without blem ishes
12-A	250-300	Small Fruits may have 1-2 spots
12-B	< 250	Very small size

Size grades for pineapple

Grade	Fruit Weight (g)
A	> 1500
B	Between 1100 - 1500
C	Between 800 - 11000
D	< 800
Baby	550 (Approx.)

Size grades for onion

Grade	Bulb dia. (cm)
Extra Large	> 6
Medium	4-6
Small	2-4

Curing: Effective operation to reduce water loss during storage from hardy vegetables viz., onion, garlic, sweet potato etc. The commodity is left in the field itself in a heap under shade for few days

Waxing: Process of applying wax on the surface of commodity by spraying, dip or immersion for improving fruit appearance and reduce water loss and shriveling during storage. Some commercial waxes are Tal-Prolong, Semper Fresh, Frutox, Waxol, Nipro Fruit wax

Packaging: Packaging of fruits is primarily undertaken to assemble the produce in convenient units for storage, marketing and distribution. There are various types of packaging material used for handling of fresh fruits and vegetables

1. *Plastic Crates / Boxes* : Used for on farm handling and transportation of fruits and vegetables. Now a days

some special crates are provided by various companies for transportation of fruits from the field to the Cold Storage. eg. Boxes provided by Adani Group for apple in Himachal Pradesh. These plastic crates are very sturdy, water resistant, ventilated and have good staking strength.

2. *Wooden Boxes* : Earlier these boxes were used for the long transportation of fruits (apple, pear, stone fruit etc.) and vegetables (tomato) but now they have become obsolete due to their various disadvantages i.e. environmental problems, injury to the fruits etc.
3. *CFB Cartons*: Corrugated Fibre Board Cortons are light in weight, easy to handle, recyclable, suitable for printing etc. These can also be made water resistant by wax coating or plastic coating. Comes in 3 ply or 5 ply types
4. *Stretch Wrapping*: Used for retail marketing of fresh produce in the form of cling plastic films. They provide modified atmosphere and improve presentation of produce.
5. *Modified Atmosphere Packaging*: Internal atmosphere is modified by some chemicals like ethysorb etc.

Pre-packaging : Packaging of the produce in consumer size units either at producing centre before transport or at terminal markets in order to reduce transportation cost by eliminating unwanted and inedible portion of fruits and vegetables. They have better presentation and eye appeal, easy handling and better storage life.

Suggested Readings

Anonymous 2005 Package of Practices of Horticultural Crops. Dr Y S Parmar University of Horticulture and Forestry, Nauni, Solan HP, India 282p.

Rajput CBS and Babu RSH 1991. Citriculture. Kalyani Pub. New Delhi p. 368.

Sharma RM and Singh RR 2000. Harvesting, postharvest handling and physiology of fruits and vegetables. *In:* Postharvest Technology of Fruits and Vegetables – Handling, Processing, Fermentation and Waste Management. Verma LR, Joshi VK (eds.) Indus Pub. Co. New Delhi, pp. 94-147.

□□□

Exercise 5

To Study Low Cost Storage Technology for Horticultural Crops

LEARNING OBJECTIVES

The students should be able to

- ❒ Learn various low cost storage techniques
- ❒ Understand the principles underlying their operation

Materials Required

Fruits/ vegetables, weighing balance, plastic crates, CFB boxes or other packaging material fruit pressure tester, hand refractometer, markers etc.

Procedure

1. Harvest the fruits
2. Subject them to various pre-treatments (as discussed in previous exercise).
3. Record their initial parameters i.e. weight, firmness, surface colour, TSS, rotting etc.

4. Place properly packed commodity in storage alongwith a check under ambient conditions
5. Record the observations regarding changes in quality during storage

Observations

Fruit Weight

Treatment	Storage period days/months						
	Initial	2	4	6	…..	……	……
Check							
ZECC							

Rotting %

Treatment	Storage period days/months						
	Initial	2	4	6	…..	……	……
Check							
ZECC							

Fruit Firmness

Treatment	Storage period days/months						
	Initial	2	4	6	…..	……	……
Check							
ZECC							

TSS

Treatment	Storage period days/months						
	Initial	2	4	6	…..	……	……
Check							
ZECC							

Surface Colour

Treatment	Storage period days/months						
	Initial	2	4	6	.....	......	......
Check							
ZECC							

For Teacher

a. The teacher shall tell the students about various low cost storage techniques for specific fruits and vegetables

b. He/she shall also explain in principles of their functioning

c. He/she shall ask the students to record various observations pertaining to the fruit quality initially and during storage.

RELEVANT INFORMATION

Some low cost storage systems

Type of low cost storage	Crops that can be stored
In situ	Root vegetables
Pits	Potato, ginger etc.
Clamps	Ginger, Potato, Cassava
Cellars	Apple, cabbage, onion, potato
Evaporative coolers - ZECC	Apple, capsicum, malta, plum, pea, pear etc.

Suggested Readings

Tripathi SN and Bhatt A 2000. Storage systems of fruits, vegetables and their products. *In:* Postharvest Technology of Fruits and Vegetables - Handling, Processing, Fermentation and Waste Management. Verma LR, Joshi VK (eds.) Indus Pub. Co. New Delhi, pp 555-596.

Exercise 6

Extraction and Preservation of Fruit Juices and Pulps

LEARNING OBJECTIVES

The students should be able to

- Extract juice/ pulp from different fruits
- Preserve different kinds of fruits in the form of juices or pulps
- Prevent spoilage of such products by microorganisms by adopting different methods of preservation and use of chemical preservatives.

Materials Required

Fruit, pulper, fruit mill, hydraulic press, screw type juice extractor, stainless steel pans, chemical preservatives, glass bottles, closures, crown corks, crown corking machine, plastic barrels.

Procedure

1. Select and weigh sound fruits/vegetables for extraction of juice/pulp.

2. Wash and peel the fruits as required.
3. Extract the juice by hot or cold process.
4. Pasteurize the juice/pulp to inactivate the enzymes and partially kill microorganisms.
5. Preserve the juice/pulp by heat processing after hot filling and hermetic sealing or by addition of permissible quantities of suitable preservatives.
6. Label the bottles properly and keep them for storage

Observations

Parameter	Apple	Orange	....
Total fruit weight (Kg)			
Juice / Pulp recovery (%) in hot extraction			
Juice / Pulp recovery (%) in cold extraction			
Peel (%)			
Seed (%)			
Residue (%)			
Temp. of pasteurization (°C)			
Duration of pasteurization			
Name of preservative used			
Conc. of preservative used (ppm)			
Temp. of hot filling (°C)			
Temp of heat processing (°C)			
Time taken for heat processing (min.)			

For Teacher

a. The teacher shall tell the students about two main methods of juice/ pulp extraction from various fruits (as detailed below).

b. He / she shall tell them which preservative (and its maximum permissible limits) are to be used for preservation.

c. He/she shall also explain the process followed for hot filling and heat processing in glass bottles.

RELEVANT INFORMATION

Methods of Juice / Pulp Extraction

1. *Hot process :* The cut/whole fruits (apple, plum, apricot, berries etc.) or vegetables (tomato, etc.) are heated over a low flame for about 5-10 min often with addition of a small quantity of water, with constant stirring to avoid charring. The boiled mixture is passed through pulper for the extraction of pulp. The *advantages* of this process are that the yield is higher and partial sterilization of juice is achieved during heating, the only disadvantage being the loss of some quantity of nutrients such as vitamins during heating.
2. *Cold Process :* The crushed fruits or berries are fed to the hydraulic press or screw type juice extractor without any pre-heating. The *disadvantages* of this method are less yield of juice or pulp, low recovery of colouring pigments, higher microbial load.

Common Methods of Fruit Preservation

a. By application of heat

Pasteurization: Preservation of fruit juices by heating them upto 100° C or below for sufficient time to kill or inactivate most but not all of the micro-organisms which can potentially cause spoilage is called pasteurization.

Processing: The fundamental method of preservation of foods by application of heat is called processing. The food is heated at or above 100° C for sufficient time to practically kill all the microorganisms present in the food.

b. Use of chemical preservatives

Sodium Benzoate is normally preferred for the products rich in anthocynins (grapes, plum, rhododendron extract etc.) while, Potassium Metabisulphite (KMS) for products containing rich quantities of carotenoid pigments (mango, apricot, lime, lemon, seabuckthorn etc.). The maximum permissible limit for use of SO_2 is 1000 ppm for fruit juices or pulps.

c. Other methods

Some other methods, although less common, are also used in some cases such as Freezing (Preservation of juices, in hermetically sealed containers by freezing at temperatures between –10 to –15 ° C), Carbonation (mixing of carbon dioxide in juices, followed by aseptic sealing), germ proof filtration (filtration of clarified juices through the filters capable of retaining bacteria, yeasts etc).

Points to remember

- Select healthy and sound fruits for juice/pulp extraction.
- Make sure that all the materials such as balances and equipments are in working condition, handling containers such as bottles pans etc are properly washed and dried, preservatives and the facilities for sealing and labeling are also available.

Suggested Readings

Lal G, Sidappa GS and Tandon GL 1998. Preservation of Fruits and Vegetables. ICAR Pub. New Delhi p488.

Exercise 7

Preparation of Beverages (RTS Drink, Squash/Cordial)

LEARNING OBJECTIVES

The students should be able to

- ❒ Understand the difference between RTS drink, squash, cordial and nectar
- ❒ Prepare different kinds of fruit beverages
- ❒ Understand the principles of preservation used for each of the prepared products

Materials Required

Fruit juice/ pulp, cane sugar, potable water, essence, citric acid, colours, preservatives, weighing balance, bottling and sealing facilities, hand refractometer, homogenizer etc.

Procedure

1. Extract the juice as discussed in previous exercise.
2. Filter the fresh or preserved juice through muslin cloth.

3. Calculate the quantities of juice, sugar, water, acid (if required), preservatives, colour, essence etc. required for preparation of a known quantity of squash as per FPO specifications.
4. Prepare the beverage as directed by instructor
5. Pack the beverage in bottles
6. Seal the bottles and label them
7. Take out one of the bottles and measure its TSS and titratable acidity.

Observations

Parameter	RTS beverage	Squash	Cordial	Nectar
Fruit juice / Pulp (qty)				
Sugar (qty)				
Acid (qty)				
Water (qty)				
Colour (qty)				
Essence (qty)				
Qty of finished product prepared				
Preservatives (Name & qty, if any)				
% of fruit part in the finished product				
TSS of finished product				
Acidity of finished product				

For teacher

a. The teacher shall tell the students standard terminology used in beverages

b. He / she shall tell them the FPO specifications for various beverages as detailed in Annexure I

c. He/she shall also explain them the stepwise process followed for the preparation of various beverages.

RELEVANT INFORMATION

Beverage: It includes all drinks fermented or unfermented, sweetened or unsweetened. Tea, coffee, coca cola, cider, maaza are all beverages.

Unfermented juice or Pure Fruit Juice: This is the natural juice pressed out of the fruit and remains practically unaltered in its composition during preparation and preservation. No additives like sugar and acid, water etc. are added.

Fruit Juice Beverage: This is the fruit juice that has been considerably altered in composition before consumption. It may be diluted before it is served as a drink eg. squashes, nectars, appetizers etc. Additives like sugar, acid, colour, essence, preservatives and water may be added.

Fermented Fruit Beverage: This is the fruit juice that has undergone alcoholic fermentation by yeast. The product contains varying amounts of alcohol. eg. grape wine. Apple cider, berry wine, plum wine, peach wine etc.

Fruit Squash: This consists essentially of strained juice containing moderate quantities (min. 25% as per FPO) of fruit pulp to which cane sugar is added for sweetening (final TSS > 40%) eg. orange squash, lemon squash, pineapple squash, mango squash etc.

Fruit Juice Cordial: This is sparkling clear sweetened fruit juice from which all the pulp and other suspended material have been completely eliminated (generally by using clarification methods) eg. lime juice cordial. This also contains moderate quantities of fruit juice (min. 25%) and added sugar to attain a TSS of min. 30% in finished product.

Sherbet: This is a clear sugar syrup which has been artificially flavoured eg. Sandal sherbet, Kewra sherbet, Rose sherbet etc. Contains no fruit juice, pulp or extract. Preferably consumed chilled after dilution in water.

Syrup: This is a clear syrup containing extract of fruits, herbs and flowers etc. eg. Brahmi Syrup. It is consumed after dilution in water.

RTS Fruit beverage/ RTS Drink: It is a beverage which is consumed directly without dilution and contains min. 10% of fruit part as well as min.10% TSS.

Fruit Juice concentrate: Fruit juice which has been concentrated by the removal of water either by heat or freezing i.e. evaporative concentration or freeze concentration. Carbonated beverages and other products are made from this eg. apple juice concentrate, pineapple juice concentrate etc.

The beverages are generally prepared from the preserved and stored juices and pulps. In case the beverage is to be prepared from fresh juice or pulp, the extraction and preservation of juices may be done as detailed in the earlier exercises. Nectar is always prepared from fresh juice and should not contain any preservatives or artificial colours or essences.

Preparation of fruit juice squash

Steps

a. Filter the juice through muslin cloth and homogenize it thoroughly if required.

b. Calculate the quantities of additives required for preparation of a known quantity of squash as per FPO specifications.

c. Add weighed sugar and acid (if any) to weighed quantities of water to prepare the sugar syrup by boiling. Remove the scum, filter the syrup and cool.

d. Mix the weighed amount of fruit juice in the syrup to dissolve thoroughly.

e. Add weighed amount of preservative, colour or essence by taking out small quantity of prepared squash in a beaker, dissolve them and adding finally to the whole lot.

f. Fill in squash bottles. Seal with PP Cap sealing machine and label before storage.

Squash should have TSS of about 40-45% and acidity of about 1.0-1.2% so that the final product obtained after dilution has a TSS of about 10-12 % and acidity of 0.2 - 0.3 %. The calculations for preparation of all the squashes shall depend on the initial TSS and acidity of the raw material i.e. pulp or juice.

Calculations

Suppose that the orange pulp has TSS of about 15°Brix and acidity of 2 % and you wish to prepare orange squash. The following calculations shall be applicable.

Pure Orange Juice	1 Litre
Expected quantity of squash to be prepared	4 Litres (as squash contains not less than 25% fruit part)
Let the TSS of final product be 45°Brix	

contd...

Table contd...

Quantity of TSS in the finished product	1800 g (45 g from 100 ml, 450 g from 1 Litre, 450 x 4 = 1800 g from 4 Litres)
TSS to be obtained form orange juice	150g (15 g from 100 ml, 15 x 10 = 150 g from 1Litre)
Quantity of sugar to be added from outside	1800-150 = 1650 g
Let the acidity of final product be 1.2 %	
Quantity of acidity in the finished product	48 g (1.2 g from 100 ml, 12 g from 1Litre, 12 x 4 = 48 g from 4 Litres)
Acidity to be obtained form orange juice	20 g (2 g from 100 ml, 2 x 10 = 20 g from 1 Litre)
Quantity of citric acid to be added from outside	48 – 20 = 28 g
Amount of preservative to be added @ 350 ppm KMS	350 mg x 4 = 1.4 g/4 Litres (The maximum permissible limits for addition of KMS in squash is 350 ppm SO_2, but we are adding 350 ppm KMS which is very less than the maximum permissible limits. KMS contains about 54% of SO_2).

Note: Practically we never get 4 Litres of squash from 1 Litre of pure juice because the weight and volume of sugar are not the same. Also, some moisture is lost by evaporation during heating of syrup. This results in reduction of volume of the finished product obtained, which is practically about 3.6 to 3.7 Litre instead of 4 Litres but all the calculations are done considering the 4 Litres of finished squash.

Preparation of RTS beverage

Steps

a. Filter the juice through muslin cloth to remove any suspended material.

b. Calculate the quantities of juice, sugar, water and acid (if required) required for preparation of a known quantity of RTS beverage as per FPO specifications.

c. Add calculated amount of sugar and acid to weighed amount of water and heat to boiling and filter through muslin cloth.

d. Add the weighed amount of juice to the boiling mixture. Heat to about 85-87°C or practical boiling.

e. Add the calculated amount of essence and colour dissolved in small amounts of the mixture (step d) towards the end of heating.

f. Hot fill the beverage into pre-sterilized glass bottles upto the brim and cork seal them immediately.

g. Sterilize the sealed bottles in boiling water for about 10-15 min and cool rapidly to room temperature.

h. Dry and label the bottles.

Calculations

Suppose that the orange juice has TSS of about 15 °Brix and acidity of 2 % and you wish to prepare RTS orange beverage. The following calculations shall be applicable.

Pure Orange Juice	1 Litre
Expected quantity of RTS beverage to be prepared	10 Litres (as RTS beverage contains not less than 10% fruit part)
Let the TSS of final product be 13°Brix	Quantity of TSS in the finished product 1300 g (13 g from 100 ml, 130 g from 1 Litre, 130x10 = 1300 g from 10 Litres)
TSS to be obtained form orange juice	150g (15 g from 100 ml, 15 x 10 = 150 g from 1Litre)

contd...

Table contd...

Quantity of sugar to be added from outside	1300-150 = 1150 g
Let the acidity of final product be 0.3 %	
Quantity of acidity in the finished product	30 g (0.3 g from 100 ml, 3 g from 1Litre, 3 x10 = 30 g from 10 Litres)
Acidity to be obtained form orange juice	20 g (2 g from 100 ml, 2 x 10 = 20 g from 1Litre)
Quantity of citric acid to be added from outside	30 – 20 = 10 g
Amount of preservative to be added @ 70 ppm KMS	70 mg x 10 = 0.7 g/10 Litres (The maximum permissible limits for addition of KMS in RTS beverage is 70 ppm SO_2, but we are adding 70 ppm KMS which is very less than the maximum permissible limits. KMS contains about 54% of SO_2). Normally you may not require addition of any preservative in RTS beverages as these are hot filled and hermatically sealed in glass bottles.

Note: Practically we never get 10 Litres of RTS beverage from 1 Litre of pure juice because the weight and volume of sugar are not the same. Also, considerable quantity of moisture is lost by evaporation during heating of beverage before hot filling. This results in reduction of volume of the finished product obtained, which is practically about 8.5-9.0 Litre instead of 10 Litres but all the calculations are done considering 10 litres of finished RTS beverage.

Preparation of fruit Nectar

Steps

a. Extract the juice as in previous exercise.

b. Filter the juice through muslin cloth to remove any suspended material.

c. Calculate the quantities of juice, sugar, water and citric acid (if required) required for preparation of a known quantity of nectar as per FPO specifications.

d. Gradually mix the weighed quantities of water into the pulp/ juice.

e. Add calculated amount of sugar and acid and mix thoroughly.

f. Strain the product through muslin cloth and heat to 85-87 °C or practical boiling.

g. Hot fill the beverage into pre-sterilized glass bottles upto the brim and cork seal them immediately.

h. Heat the sealed bottles in boiling water for about 10-15 min and cool rapidly to room temperature.

i. Dry and label the bottles.

Preparation of Lime Juice Cordial

Cordial is a product that should essentially be prepared from a clarified juice so that the finished product does not have any suspended material. Therefore the clarification of lime juice is must before starting the preparation of product.

Clarification of juice: Various clarifying agents i.e. gelatin, egg albumin, casein, mixture of gelatin and tannin are used for clarification of juices. These clarifying agents are added to the juice alongwith KMS and stored in glass carboys or in upright wooden barrels for about 2-3 months to get a clear juice, which is decanted or siphoned for use. This process is slow and time consuming.

Enzymes, such as pectinase and cellulase are used commercially for the clarification of juice. The juices should be treated with these enzymes @ 0.2 % for about 2 hrs at 50 ± 2° C followed by filtration under suction through filter press to get a sparkling clear juice in a short time.

Steps

1. Calculate the quantities of clarified juice, sugar and water required for preparation of a known quantity of lime juice cordial as per FPO specifications.
2. Prepare syrup of 40°Brix. Cool and filter through thick muslin cloth.
3. Mix the clarified juice and filtered syrup and add edible colour, preservatives etc.
4. Fill product in pre sterilized glass bottles leaving about 2.5 - 3.0 cm head space and are sealed.
5. Dry and label the bottles.

Suggested Readings

Lal G, Sidappa GS and Tandon GL 1998. Preservation of Fruits and Vegetables. ICAR Pub. New Delhi p488.

Sharma SK, Sharma PC and Kaushal BBL 2001. Effect of storage temperature and folds of concentration on quality characteristics of Galgal (Citrus pseudolimon. Tan.) juice concentrates. J Fd. Sci. Technol. 38(6): 553-556.

Vaidya D and Vaidya M 2000. Fruit juices and juice beverages. *In:* Postharvest Technology of Fruits and Vegetables – Handling, Processing, Fermentation and Waste Management. Verma LR, Joshi VK (eds.) Indus Pub. Co. New Delhi, pp 681-719.

❑❑❑

Exercise 8

Preparation of Fruit Juice Concentrates

LEARNING OBJECTIVES

The students should be able to

- Understand the different methods used for the preparation of fruit juice concentrates.
- Understand the stepwise process for preparation of fruit juice concentrates
- Understand the factors responsible for causing loss of nutritional and sensory quality attributes.

Materials Required

Fruits juice, enzymes (pectinase, cellulase, pectinase CCM etc.), filter press, buchner filter apparatus, water bath, filter paper pulp, buchhi type rotary vacuum evaporator, freeze concentrator, membrane concentrator (whichever available), vacuum grease.

Procedure

1. Extract the juice as discussed in earlier exercises. Take a known quantity of juice in a conical flask.
2. Add enzymes i.e. pectinase @ 0.2% to the juice and keep in a water bath at temperature of 40-50° C for 2 hrs.
3. Shake the flasks occasionally for uniform mixing of enzyme all throughout the flask during the enzymatic treatment.
4. Leave the flasks undisturbed for 10-12 hrs for allowing the insoluble solids to settle at the bottom of the flask.
5. Siphon the clear juice from the top of the flask.
6. Pass this siphoned juice through filter press or through a 0.4-0.5 cm thick bed of filter paper pulp in a Buchner filter apparatus or filter press.
7. Record the volume of the clarified juice obtained. Calculate the wastage.
8. Concentrate the juice in rotary vacuum evaporator at 50±2° C temperature and at 28-30″ Hg vacuum to a TSS of about 50 to 70° Brix. Record the exact TSS.
9. Record the volume of condensate obtained in the round bottom flask of the buchhi apparatus.
10. Calculate the folds of concentration

Observations

Parameter	Observations
Name of fruit juice	
Volume of juice taken for clarification	
TSS of the juice taken for clarification	
Concentration of enzyme used	

contd...

Volume of juice obtained after clarification
TSS of juice obtained after clarification
Volume of concentrate obtained
TSS of concentrate
Time taken for concentration
Volume of condensate obtained
TSS of condensate obtained
Temperature of water bath during concentration
Vacuum level during concentration
Loss in Vitamin C content after vacuum evaporation

11. If the concentration is being done in freeze concentrator record the following observations
12. If freeze concentrator is not available, teacher can also make the students understand the concept by making use of the freezer camber of a refrigerator.
13. Place a known quantity of juice in the freezer of a laboratory refrigerator for about 2-3 hrs
14. Pass the material through a sieve.
15. Record the weight and volume and TSS of the filtrate obtained.
16. Measure the quantity of material left over on the sieve i.e. ice crystals with entrapped juice solids.

Parameter	Observations
Name of fruit juice	
Volume of juice taken for freeze concentration	
Initial TSS of juice taken for freeze concentration	
TSS of concentrate	
Volume of concentrate obtained	
Quantity of juice wasted by entrapment in ice crystals	
Loss in vitamin C after freeze concentration	

The method of freeze concentration as detailed at pt. 11-16 above is not commercial one, but can be demonstrated to the students in order to make them understand the basic concept of freeze concentration.

For Teacher

a. The teacher shall tell the students the FPO specifications for juice concentrates
b. He/she shall also explain them the stepwise process followed for the enzymatic clarification of juice and its importance before preparation of concentrates.
c. He/she shall also explain them the stepwise process followed for the preparation of juice concentrates by evaporation as well by freeze concentration.
d. He/she shall ask the students to calculate the folds of concentration.

RELEVANT INFORMATION

Juice concentration is a process of separation of water from the solids of the juice by making use of the basic three methods i.e. evaporation, freezing and membrane technology. Pulps and juices are the basic material used for preparation of various types of processed products. However, preservation of single strength juice is not at all economical due to the presence of large quantities (75-90%) of water in different juices. Concentration of juices not only provides microbial stability, but also reduce the costs of packaging, transportation and storage of juices by reducing the bulk of the material.

Methods of Concentration

1. *Evaporative concentration*: This method involves the removal of water from the juice by converting the

liquid into vapours at low temperature (about 50° C) under vacuum in order to retain maximum nutritive and sensory quality. Here we use the difference in boiling point of solute and solvent as the basis for separation. This method is not very expensive and is practically used at commercial levels for the preparation of concentrates. Some loss to the nutritional and sensory quality of juice occurs during the process of concentration but we can achieve very high TSS (> 70°Brix) of concentrates by this method.

2. *Freeze concentration*: This process involves the separation of water from the juice by freezing the liquid into ice and then separation of ice crystals. This process uses the difference in freezing point of solute and solvent as the basis for the separation. The equipment is very costly and loss of juice solids may occur due to entrapment in ice crystals. Quality of concentrate is superior than that obtained by first method. We can not achieve very high TSS of concentrate by this method.

3. *Membrane concentration*: This process involves the separation of water from the solids by passing the juice under pressure through a membrane which permits only smaller particles to pass through it. Bigger sized particles are retained on the top. It uses the difference of particle size as the basis of separation. Quality of concentrate is superior than that obtained by first method. We can not achieve very high TSS of concentrate by this method.

Clarification of juices: For achieving high TSS of the concentrates, clarification of juices by enzymatic methods is very important. Unclarified juice when concentrated may lead to formation gel at high concentrations and will not be free flowing.

Suggested Readings

Sharma SK, Sharma PC and Kaushal BBL 2001. Effect of storage temperature and folds of concentration on quality characteristics of Galgal (*Citrus pseudolimon.* Tan.) juice concentrates. J Fd. Sci. Technol. 38(6): 553-556.

❑❑❑

Exercise 9

Preparation of Jam

LEARNING OBJECTIVES

The students should be able to

- Understand the difference between jam, jelly & marmalade
- Prepare jam from single or mixed fruits or preserved pulps thereof
- Understand the principles of preservation used for the prepared products

Materials Required

Fruit pulp, cane sugar, potable water, essence, citric acid, colours, preservatives, weighing balance, glass jars (500 ml capacity), jar closures, hand refractometer, steel pan, ladle etc.

Procedure

1. Select sound and fresh fruits, free from any blemishes. Generally a mixture of ripe and unripe fruits should be taken as the ripe fruits impart good flavour and the unripe fruits are rich in pectin, required for a good set.

2. Wash the fruits thoroughly. Cut into 4-8 pieces depending upon the fruit and size
3. Boil the fruit pieces for about 20-25 min in boiling water and pass through pulper to get a good pulp.
4. Weigh known quantity of fresh or preserved pulp and add about equal quantity of sugar (preferably in 45:55 ratio)
5. Cook the mixture on gas flame by occasional stirring.
6. When the mixture becomes thick, perform sheet test.
7. Citric acid is added generally towards the end of cooking. Addition of all the acid in the beginning may result in stickiness and poor set.
8. Remove from the flame when the mixture gives a clear sheet
9. Flavours, colours etc may also be added towards the end of boiling.
10. Fill the jam in glass jars (preferably pre-warmed in hot air).
11. Close the jars with lug caps and allow to cool overnight.
12. Label and store in a cool place.

Observations

Parameter	Observations
Fruit / Pulp (qty)	
Sugar (qty)	
Acid (qty)	
Pectin (qty)	
Water (qty)	
Colour (qty)	
Essence (qty)	
Qty of finished product obtained	
Preservatives (Name & qty, if any)	
TSS of finished product	
Acidity of finished product	
Method of testing end point	

For Teacher

a. The teacher shall tell the students the difference between jam, jelly and marmalade

b. He / she shall tell them the FPO specifications for various products as detailed in Annexure I

c. He/she shall also explain them the stepwise process followed for the preparation of jam.

d. The different methods for determining the end point should also be told.

RELEVANT INFORMATION

Jam : It is a product prepared by boiling to a suitable consistency, fruit pulp with sufficient quantity of sugar so as to become firm enough to hold the tissues in position. About 45 parts of pulp are used with every 55 parts of sugar. The finished product should not contain less than 68 per cent total soluble solids.

Jelly : Jelly is prepared by boiling the fruit extract with sugar in the presence of appropriate amounts of acid and pectin so as to set into a clear gel. A perfect jelly is transparent, well set, but not too stiff and should have original flavour of the fruit. When cut, it should retain its shape and show a smooth cut surface.

Marmalade : It is a fruit jelly in which the slices of the fruit or the peel are suspended. Mostly they are prepared from citrus fruits like oranges, lemons etc.

Principle of Preservation in Jam, Jelly and Marmalade : Sugar under high concentration (>66%), acts as a preservative. The moisture present in jam becomes bound and is not free for the growth of micro organisms. The water activity is reduced.

Difference between Jam and Jelly

Jam	Jelly
1. Prepared from fruit pulp	1. Prepared from clear fruit extract
2. Not transparent	2. It is transparent
3. Min. TSS 68% as per FPO specifications	3. Min. TSS 65% as per FPO specifications
4. Flavours and colours are generally added	4. Should have the original flavour of the fruit.

Determination of End Point

Although the end point can be judged easily with the increase in experience, but for the beginners the following methods can be used.

1. *Drop Test*: This method can be used where no facility for any type of measurement is available. A small quantity of product is taken in a spoon during cooking and is allowed to air cool for some time. One drop of the air cooled jam is put into a glass of water. If the drop settles down at the bottom of the glass without disintegrating, the product is ready. If it disintegrates and splits apart, it may be cooked for some more time.

2. *Boiling point:* Jam is assumed to be ready when the temperature of the boiling mixture reaches 106° C at sea level. The temperature may be a little lesser at high altitudes.

3. *TSS:* At end point the TSS of the jam is above 68°Brix. Care should be taken to cool the drop of material which is put on the prism of refractometer. High temperatures may lead to condensation of vapours and thus giving incorrect readings.

4. *Weight Test:* At the end point, the weight of the jam is about 1.5 times the weight of the added sugar. For using this method the weights of the boiling pan and sugar are to be noted down before starting cooking. The boiling mixture is weighed alongwith the pan for determination of end point.

Calculations

Suppose that the apple pulp has TSS of about 15 °Brix and acidity of 1.2 % and you wish to prepare Apple Jam. The following calculations shall be applicable.

Apple Pulp	2 Kg
Sugar to be added	2.4 Kg (Pulp and Sugar are added in ratio of 45:55)
Expected quantity of Jam to be prepared	3.6 Kg (Weight of final Jam is about 1.5 times the weight of sugar added)
Suppose the acidity of the finished products is 1 %.	
Quantity of acid required in the finished product	36g
Quantity of acid contributed by the pulp	1.2 x 10 x 2 = 24 g
Quantity of citric acid to be added	36 – 24 = 12 g

Note: Some weight loss in the finished product may be due to evaporation and some wastage in the container or cooking pan. This is significant when product is prepared in small quantities. The weight loss is very sharp when product is concentrated beyond 70° Brix).

Suggested Readings

Bhardwaj JC 2000. Jam, jellies and marmalades. *In:* Postharvest Technology of Fruits and Vegetables – Handling, Processing, Fermentation and Waste Management. Verma LR, Joshi VK (eds.) Indus Pub. Co. New Delhi, pp 597-636.

Lal G, Sidappa GS and Tandon GL 1998. Preservation of Fruits and Vegetables. ICAR Pub. New Delhi p488.

Srivastava RP and Kumar S 2002. Fruit and Vegetable Preservation - Principles and Practices. International Book Distributing Co., Lucknow. p474.

❐❐❐

Exercise 10

Preparation of Jelly and Marmalade

LEARNING OBJECTIVES

The students should be able to

- ❒ Understand the difference between jam, jelly and marmalade.
- ❒ Prepare jelly and marmalade.
- ❒ Understand the principles of preservation used for the prepared products.

Materials Required

Fruit extract, cane sugar, potable water, essence, citric acid, colours, preservatives, weighing balance, glass jars (500 ml capacity), jar closures, hand refractometer, steel pan, ladle etc.

Procedure for Jelly

1. Select sound and fresh fruits, free from any blemishes. Generally a mixture of ripe and unripe fruits should

be taken as the ripe fruits impart good flavour and the unripe fruits are rich in pectin, required for a good set.

2. Wash the fruits thoroughly. Cut into 4-8 pieces depending upon the fruit and size
3. Boil the fruit pieces in water (about ½ to an equal weight of fruits for fruits like apple and 2-3 times the weight of sliced fruits in case of citrus fruits) for about 30 min. with occasional stirring. Strain the boiled mixture through coarse muslin cloth to get a clear pectin extract
4. A second extract may also be taken by boiling the same mass again and straining.
5. Add sugar (usually 3/4th to one part of the extract or in 45:55 ratio) to the pectin extract and boil in a stainless steel pan. Add required quantities of acid towards the end of boiling.
6. Remove the scum rising on the top of boiling mass occasionally.
7. To avoid charring, occasional stirring is also done.
8. To avoid destruction of pectin the cooking should be completed fast
9. Determine the end point of jelly as detailed ahead.
10. Fill in hot dry glass jars and cool rapidly
11. Pour a thin layer of hot molten paraffin wax seal to avoid moulding.
12. Label and store in a cool place.

Procedure for Marmalade

1. The procedure is similar as in case of jelly except for the following points

2. Boil the sliced fruits in 2-3 times its weight in water for about 45-60 min. for extraction of pectin. Keep the extract overnight for clarification and decant or siphon to get a clear extract or pass through filter press.
3. For preparing peel shreds, the peel is also cut into shreds (18 – 25 mm long and about 1-2 mm thick). In order to avoid hardening of peel on boiling with sugar during cooking of jelly, the shredded peel is boiled for 10-15 min. in water, or the peel is boiled in 0.25 % solution of sodium carbonate or 0.1 % ammonia or the peel is softened by autoclaving at 115-120° C for some time.
4. The shreds are added @ 60-70 g/kg of original extract.
5. Occasional stirring is done in order to distribute the shreds evenly in the marmalade.
6. Perform the sheet test for determination of end point.
7. Add a small quantity of peel oil at the time of filling to make for the loss of flavour during cooking.
8. Fill the marmalade in clean glass jars. Allow to set overnight and seal the following day with hot paraffin wax.
9. In order to prevent the darkening of marmalade during storage, KMS @ 0.01% may be added during cooling.

Observations

Parameter	Jelly	Marmalade
Fruit Pulp (qty)		
Water added for extraction (qty)		
Extract obtained (qty)		
Sugar (qty)		
Acid (qty)		

contd...

Water (qty)
Colour (qty)
Essence (qty)
Qty of finished product obtained
Preservatives (Name & qty, if any)
TSS of finished product
Acidity of finished product
Method of testing end point
Problems faced during preparation of extract and remedial action taken
Problems faced during preparation of product and remedial action taken
Problems faced during jar filling and packing and remedial action taken

For Teacher

a. The teacher shall tell students the FPO specifications for various products as detailed in Annexure I

b. He/she shall also explain them the stepwise process followed for the preparation of jelly and marmalade.

c. The different methods for determining the end point should also be told.

d. He/she shall also tell the precautions / remedies for various common problems faced during the process of product preparation and packing.

RELEVANT INFORMATION

The end point in case of jelly and marmalade can be judged by the following methods.

1. *Boiling point:* Sugar solution containing 65% sugar boils at 105° C at sea level. The same shall hold true for jelly, but temperature corrections are necessary depending upon the altitude of the place. Generally,

the end point of jelly is 3-4° C higher than the boiling point of water at that place.

2. *Sheet or ladle test:* Take some boiling jelly in a ladle and cool it slightly and allow it to drop into the pan. If it drops like syrup, the mixture requires more cooking and if it falls like a sheet or flake, the end point has been reached.

3. *Cold plate Test:* Take a drop of boiling mixture and place it on a cold inverted steel plate. Cool it rapidly and push it slightly with a finger. If the jelly is ready, the cooled drop will crinkle and come all at once without breaking.

4. *Weight Test:* At the end point, the weight of the jelly should be 1.5 times the weight of the added sugar. For using this method the weights of the boiling pan and sugar are to be noted down before starting cooking. The boiling mixture is weighed alongwith the pan for determination of end point.

Suggested Readings

Bhardwaj JC 2000. Jam, jellies and marmalades. *In:* Postharvest Technology of Fruits and Vegetables - Handling, Processing, Fermentation and Waste Management. Verma LR, Joshi VK (eds.) Indus Pub. Co. New Delhi, pp 597-636.

Lal G, Sidappa GS and Tandon GL 1998. Preservation of Fruits and Vegetables. ICAR Pub. New Delhi p488.

Srivastava RP and Kumar S 2002. Fruit and Vegetable Preservation - Principles and Practices. International Book Distributing Co., Lucknow. 474p.

❒❒❒

Exercise 11

Preparation of Lime Chilli-Ginger-Pickle

LEARNING OBJECTIVES

The students should be able to

- ❒ Prepare lime-chilli-ginger pickle
- ❒ Understand the principles of preservation used for pickles

Materials Required

Lime fruits, green chillies, ginger salt (common), stainless steel pans, kinives, weighing balance, spices etc.

Procedure

1. Select sound and mature fruits of lime, mature but green chillies having good size and ginger rhizomes.
2. Wash the lime fruits, ginger and chillies properly in running cold water.
3. Cut the lime fruits into pieces / halves.

4. Remove the stalks of the chillies and cut into longitudinal slices. Peel ginger and cut into long thin slices.
5. Put the lime pieces, chilli and ginger slices in a wide mouthed pre-sterilized glass jar
6. Add common salt @ 250g / Kg of material.
7. Squeeze some of the juice over the slices
8. Stir the mass well mixing thoroughly
9. In case the juice does not cover the material completely, add pre-boiled and cooled 5% solution of citric acid.
10. Keep the jar in Sun for about a week with occasional shaking of the material.
11. When the lime pieces start turning light brown, the pickle is ready for consumption.

Observations

Parameter	Observations
Lime Fruit (qty)	
Chilli (qty)	
Ginger (qty)	
Salt (qty)	
Citric Acid if required (qty)	
Spices added (if any)	
Colour of material before keeping in Sun	
Days taken for turning the colour brownish	
Days taken for softening lime pieces	
Daily Temperature of the liquid portion	
Preservatives added (if any)	

For Teacher

a. The teacher shall tell the students the FPO specifications for different kinds of pickles as detailed in Annexure I

b. He/she shall also explain them the stepwise process followed for the preparation of lime-chilli-ginger pickle

c. He/she shall also tell the principles of preservation used in different pickles

RELEVANT INFORMATION

Pickling : The process of preservation of foods in common salt or vinegar is called pickling and the products thus obtained are called pickles. Spices and vegetable oils may also be added for imparting good taste, flavour and for their preservative properties.

Pickle is an edible product preserved and flavoured in a solution of common salt and vinegar with added spices and oil.

Principles of Preservation

- Conversion of fermentable carbohydrates into organic acids.
- Addition of sufficient amount of sugar, vinegar and other ingredients to the fully cured and packed product.
- Antimicrobial activity of organic acids, oils, spices etc.

Pickling Process
2 stages

1. Curing or fermentation with dry salting or fermentation in brine or salting without fermentation
2. Finishing and packaging

The lactic acid bacteria are salt tolerant and they can flourish in a brine of about 8-10% salt concentration. The various fruits and vegetables which undergo lactic aicd fermentation are cabbage, cauliflower, broccoli, carrot, turnips, radish, cucumber, olive, garlic, apples, pear, immature mangoes, immature plums, lemons, banana etc.

Lactic Acid Fermentation

Homofermentative bacteria
Carbohydrates ----------------→ Lactic Acid

Heterofermentative bacteria
Carbohydrates ----------------→Lactic Acid, Acetic acid, CO_2, Ethanol

Gherkins: Small sized cucumbers brined for pickling.

Sauerkraut: Fermented cabbage: It is a sound, clean product having specific flavour obtained by full fermentation, predominantly lactic acid fermentation, of properly prepared and shredded cabbage in the presence of 2 to 3% salt. It contains upon complete fermentation, not less than one and a half per cent lactic acid.

FPO Specifications

1.	Pickles in vinegar	Not less than 2% acidity in fluid portion as citric acid.
2.	Pickles in citrus or brine	Not less than 12% salt
		not less than 1.2% acid if the pickle is made in citrus juice
3.	Oil Pickles	Any edible vegetable oil like mustard, olive oil etc should be used.

Suggested Readings

FPO 1991. The Fruit Products Order 1955. Ministry of Food Processing Industries, Government of India, 49p.

Joshi VK and Bhatt A 2000. Pickles-technology of preparation. *In:* Postharvest Technology of Fruits and Vegetables - Handling, Processing, Fermentation and Waste Management. Verma LR, Joshi VK (eds.) Indus Pub. Co. New Delhi, pp 777-820.

Lal G, Sidappa GS and Tandon GL 1998. Preservation of Fruits and Vegetables. ICAR Pub. New Delhi 488p.

Srivastava RP and Kumar S 2002. Fruit and Vegetable Preservation - Principles and Practices. International Book Distributing Co., Lucknow. 474p.

❑❑❑

Exercise 12

Preparation of Fruit Chutney

LEARNING OBJECTIVES

The students should be able to

- Prepare fruit chutney
- Understand the principles of preservation used for preparing chutney
- Learn the FPO specifications for chutney

Materials Required

Fruits or fruit pulp, steel pans, weighing balance, refractometer, sugar, salt, spices, acetic acid, citric acid etc.

Procedure

1. Extract fruit pulp as told in earlier exercises.
2. Take a known quantity of freshly extracted or preserved pulp.

3. Weigh sugar (about half the weigh of pulp).
4. Heat pulp and sugar over a gas flame.
5. Add spices, chillies, salt, citric acid (if required) etc. and cook
6. Record the TSS to judge the end point. It should be more than 50° Brix.
7. Add measured quantities of acetic acid and sodium benzoate at the end.
8. Cool and pack in pre-washed and sterilized glass jars
9. Label and store.

Observations

Parameter	Observations
Name of fruit or vegetable	
Weight of pulp taken	
TSS of pulp	
Acidity of pulp	
Quantity of sugar to be added	
Quantity of all the spices, salt, acetic acid and citric acid to be added	
Amount of preservative (Sodium Benzoate) reqd.	
Time taken for preparation	
TSS of the finished product	
Weight of the finished product obtained	

For Teacher

a. The teacher shall tell the students the FPO specifications for fruit chutney
b. He/she shall also explain them the stepwise process followed for the preparation of fruit chutney
c. He/she shall also tell the principles of preservation used in chutney

RELEVANT INFORMATION

Chutney: It is a product prepared by cooking the fruit pulp with added salt, sugar, spices, acetic acid and / or dry fruits to a suitable consistency. Sugar, salt, spices, acetic acid all act as partial preservatives. According to the FPO fruit chutney should have a minimum of 50% TSS and 40% fruit part. Preservatives may also be added in chutney for increasing its storability. Generally no artificial colour and flavour are required to be added.

A general recipe for the preparation of fruit chutney

Fruit pulp (TSS 10-15° Brix and acidity about 1%)	5 Kg
Sugar	2.4-2.6 Kg
Salt	140-150g
Red chilli powder (according to taste)	15-25 g
Black pepper	35-40 g
Cardamom	70-80g
Cumin	40-50 g
Ginger chopped	350g
Citric acid	15-20g
Acetic acid	15-20 ml
Sodium Benzoate	1 g

Suggested Readings

FPO 1991. The Fruit Products Order 1955. Ministry of Food Processing Industries, Government of India, 49p.

Lal G, Sidappa GS and Tandon GL 1998. Preservation of Fruits and Vegetables. ICAR Pub. New Delhi 488p.

Srivastava RP and Kumar S 2002. Fruit and Vegetable Preservation - Principles and Practices. International Book Distributing Co., Lucknow. 474p.

❑❑❑

Exercise 13

Preparation of Tomato Products (Puree, Ketchup)

LEARNING OBJECTIVES

The students should be able to

- ❐ Understand different tomato products
- ❐ Prepare tomato products
- ❐ Understand the principles of preservation used for tomato products
- ❐ Learn the FPO specifications for various tomato products

Materials Required

Tomato fruits, pulper, steel pans, weighing balance, refractometer, sugar, salt, spices, acetic acid, citric acid, muslin cloth etc.

Procedure

I. Puree

1. Extract fruit pulp by Hot process as told by the teacher.
2. Record the TSS of the pulp.
3. Concentrate the pulp by heating to 12° Brix (heavy puree)
4. Hot fill the boiling puree into pre-sterilized glass bottles (200 ml cap.)
5. Seal the bottles by crown corking
6. Heat process in boiling water in a big aluminium pan for 15-20 min.
7. Cool the bottles to room temperature as quickly as possible.
8. Label and store at a dry place.

II. Ketchup

1. Extract fruit pulp by Hot process as told by the teacher.
2. Record the TSS of the pulp.
3. Heat the pulp in a steel pan
4. Add spices salt etc. during boiling as per recipe. Add spice extract or directly boil spices in a muslin cloth bag properly tied, to avoid any pilferage of spices directly into the sauce/ketchup.
5. Record the TSS at end point i.e. > 25% (generally TSS more than 28 is kept in good ketchups).
6. Sodium benzoate @ 750 ppm may be added as preservative.
7. Hot fill in pre-sterilized glass bottles

8. Cool the bottles to room temperature as quickly as possible.
9. Label and store at a dry place.

Observations

Parameter	Observations
Weight of tomato fruits	
Weight of pulp obtained	
TSS of pulp	
Acidity of pulp	
Method of addition of spices	
TSS of puree / ketchup at end point	
Amount of preservative (Sodium Benzoate) reqd.	
Weight of the finished product obtained	

For Teacher

a. The teacher shall tell the students the FPO specifications for various tomato products.
b. He/she shall also explain them the stepwise process followed for the preparation of puree and ketchup
c. He/she shall also tell the principles of preservation used in these tomato products

RELEVANT INFORMATION

Selection of Fruit

- Plant ripened red tomatoes should be selected. Yellow and green fruits on processing yield dull colour and turn brown due to oxidation.
- Never use iron equipment for processing. Lycopene oxidizes and turns brown when comes in contact with

iron. It also forms black compounds with the tannins present in tomatoes or the spices which are used for product preparation.

- Avoid prolonged heating.

Product Specifications

Product	Min TSS (° Brix)	Min acidity
Juice	5	
Soup	7	
Puree		
Medium	9	
Heavy	12	
Paste	25	
Ketchup	25	1.0
Sauces mixed	15	1.2

Characteristics of Juice

- Should be deep red in colour
- Should possess the characteristic taste and flavour of tomato
- Acidity as citric acid should be about 0.4%
- Should be uniform in quality

Crushing: Fruits are cut into 4-6 pieces and crushed in a fruit mill.

Pulping

A. Hot Pulping / Hot Process

Crushed tomatoes are boiled in their own juice in stainless steel pans for 3-5 min.

Advantages

1. The tendency of juice to separate into liquid and pulp can be overcome due to release of pectin into the juice and inactivation of pectase enzymes
2. Partial sterilization of juice is achieved due to reduction in microbial load.
3. Red colour is also released during cooking
4. Higher yields.

B. Cold Pulping / Cold Process

Tomatoes are crushed and passed through pulper without boiling.

Disadvantages

1. Less yield
2. Incorporation of air into juice / pulp so encourages oxidation
3. Less red colour released into juice
4. Chances of microbial spoilage are more.
5. Separation of juice into watery portion and pulp.

TOMATO JUICE

Juice Extraction: It is done through different types of juice extractors, namely Continuous Spiral Press and Cyclone or pulper.

Total Solids: Generally the total solids are 5.66%. Specific gravity 1.0240 at 20° C. About 0.4-0.6% salt and about 1% sugar can also be added.

Packaging: Done in glass bottles or cans. Before packaging, the juice is generally homogenized to retard separation of liquid from the pulp. Juice is heated at 88° C and filled hot into glass bottles followed by hermetically sealing and sterilization in boiling water (100°C) for 30 min.

TOMATO PUREE

The commercial pulp without skin and seeds, with or without added salt and containing not less than 8.37% salt free tomato solids is "medium tomato puree". It is further concentrated to 12% solids to form "Heavy Tomato Puree". Pulp is extracted and concentrated in open cookers or vacuum pans and packaging is done in glass bottles or cans.

TOMATO PASTE

A concentrated tomato juice or pulp without skin and seeds, containing not less than 25% of tomato solids is known as tomato paste.

CONCENTRATED TOMATO PASTE

Contains 33% tomato solids. Common salt, basil leaf or sweet oil of basil leaves may also be added.

TOMATO COCKTAIL

Product is prepared by adding spices including chilli, cloves, cardamom, coriander alongwith vinegar and common salt to the tomato juice. Lemon or lime juice etc may also be added in different proportions to suit the palate.

TOMATO KETCHUP

It is a product made by concentrating tomato juice or pulp without seeds and skin. Spices, salt, vinegar, onion, garlic etc. may be added so that it contains not less than 12 % tomato solids and generally 28% or more total solids (not less than 25% as per FPO specifications).

Addition of Ingredients

Spices: Good quality spices should be added in proper proportions.

a. *Bag method:* Spices / powders are tied loosely in a muslin cloth bag and placed in boiling tomato juice.

b. *Spice extracts:* Spices are boiled for long time to prepare extract which is added to the juice during ketchup preparations.

Sugar: 1/3rd of sugar is added initially in order to fix the red colour in tomato. Rest of the sugar is added a little before the ketchup is ready. Sugar is added @ 6-8 % of the tomato juice depending upon the recipe (10-26%) in finished product.

Salt: Salt is added towards the end of boiling @ 0.8% to 1.0% so that the finished product contains 1.3-3.4% salt.

Vinegar: Ketchups contain 1.25-1.50 % acetic acid. The vinegar should also be added towards the end when the ketchup has thickened sufficiently. Otherwise the acetic acid may get volatized leaving the ketchup deficient in acid as well as flavour.

Thickening agents: Pectin is usually added @ 0.1-0.2% by weight of the finished product.

Concentration: Commercial ketchups have generally 28-30% total solids and may be as high as 37%. The increase in solids increases its keeping quality.

Preservatives: Sodium Benzoate @ 750 ppm may be added in finished product.

Bottling: Ketchup is filled in bottles at 88° C and are pasteurized for 30-35 min. in hot water at 85-88° C after corking. It is preferable to add 250ppm sodium benzoate and then pasteurize the product.

Difference between ketchup and sauce

Ketchup	Sauce
Prepared from tomato only	Prepared from tomato as well as other fruits such as pumpkin, chilli etc.
Min. TSS 25%	Min. TSS 15%
Min. acidity 1.0%	Min. acidity 1.2%
Thicker in consistency	Thinner in consistency
Costly	Cheap as compared to ketchup.

Suggested Readings

FPO 1991. The Fruit Products Order 1955. Ministry of Food Processing Industries, Government of India, 49p.

Lal G, Sidappa GS and Tandon GL 1998. Preservation of Fruits and Vegetables. ICAR Pub. New Delhi 488p.

Srivastava RP and Kumar S 2002. Fruit and Vegetable Preservation - Principles and Practices. International Book Distributing Co., Lucknow. 474p.

❑❑❑

Exercise 14

Canning of Fruits and Vegetables

LEARNING OBJECTIVES

The students should be able to

- Understand the principles of preservation used for canning of fruits and vegetables
- Differentiate the canning requirements for fruits and vegetables
- Perform proper selection of fruits and vegetables for canning
- Understand and experience the steps involved in filling, exhausting, sealing etc.

Materials Required

Fruits or vegetables, stainless steel knife, steel pans, Can Body Reformer, Flanger, Double seamer, weighing balance, Refractometer / Salometer, sugar or salt etc.

Procedure

1. Wash the selected fruit or vegetable (as the case may be)

2. Cut into pieces or slices (the size and shape may depend on the fruit/ vegetable but it should not be too large or too small)
3. Blanch the material in boiling water for 1-2 min for effective inactivation of enzymes.
4. Prepare the sugar syrup of strength 40° Brix or brine of 1-3 % strength.
5. Prepare the can body using a can body reformer and flange the ends in a flanger.
6. Seal the base plate using first and second roller operations of the double seamer.
7. Transfer the prepared fruits / vegetables into the cans.
8. Pour in hot syrup or brine into the cans.
9. Place the lid over the top of filled can. (follow the first roller operation of the double seamer, if the cans are OTS cans)
10. Exhaust the cans in boiling water for about 10 min. to exclude any dissolved air out of the tissues or syrup/ brine.
11. When the temperature of the centre of can reaches 79-80 °C for fruits or 82 °C for vegetables, remove the cans from the exhaust box and seal them properly (follow the second roller operation of the double seamer, if the cans are OTS can).
12. Process the sealed cans at 100°C and 115-121°C for fruits and vegetables respectively for 15-30 min in case of fruits and 25-45 min. in case of vegetables.
13. Cool the cans immediately in cold water and air dry them at room temperature.
14. Label and store in a cool dry place.

Observations

Parameter	Observations
Name of fruit or vegetable	
Weight of fruits or vegetables taken	
Peeling performed or not (if yes, how?)	
Size of pieces	
Time taken for blanching	
Weight after blanching	
Strength of syrup or brine	
Amount of sugar/ salt and water taken for preparation of syrup or brine of desired strength	
Weight of slices added per can	
Weight of syrup/ brine added per can	
Temperature of exhausting	
Temperature and duration of heat processing	
Total can weight after sealing	

For Teacher

a. The teacher shall tell the students the FPO specifications for preparation of canned fruits and vegetables

b. He/she shall also explain them the stepwise process followed for the preparation of canned fruits or vegetables

c. He/she shall also tell the principles of preservation used in canning

RELEVANT INFORMATION

Canning: Canning is the preservation of foods in sealed containers. Fruits are generally canned in sugar syrups and vegetables in brine solutions. Most fruits are processed in boiling water i.e. 100° C (due to their high acidity), while vegetables except the likes of tomato are processed at 115 -121°C (due to their lower acidity).

Principle: Sterilization of food in hermetically sealed containers by application of heat of variable intensities and for different duration of time.

Testing of Cans

a. *Ringing sound*: A clear ringing sound indicates perfect seal while dull and hollow sound indicates imperfect seal

b. Concavity of the lid also gives and idea about the degree of vacuum inside the can

c. Vacuum gauge

d. *Flip tester*: By placing the can inside a glass chamber and evacuating it.

Suggested Readings

Lal G, Sidappa GS and Tandon GL 1998. Preservation of Fruits and Vegetables. ICAR Pub. New Delhi 488p.

Sandhu KS and Bakshi AK 2000. Commercial cannig of fruits and vegetables. *In:* Postharvest Technology of Fruits and Vegetables - Handling, Processing, Fermentation and Waste Management. Verma LR, Joshi VK (eds.) Indus Pub. Co. New Delhi, pp 637-680.

Srivastava RP and Kumar S 2002. Fruit and Vegetable Preservation - Principles and Practices. International Book Distributing Co., Lucknow 474p.

❏❏❏

Exercise 15

Cut Out Analysis of Canned Products

LEARNING OBJECTIVES

The students should be able to

- ❐ Understand cut out analysis
- ❐ Perform cut out analysis
- ❐ Conclude whether the product was prepared as per FPO specifications

Materials Required

Canned product, weighing balance, can cutter, vacuum/pressure guage, seam checking guage, Refractometer / Salometer etc.

Procedure

1. Take a canned product say canned peach.
2. Record the weight of the can (A)

3. Record the information marked on the label i.e net weight, date of packing, etc
4. Observe the can externally for the concavity of the lid.
5. Record the vacuum of the can by using a vacuum guage
6. Cut the lid of the can with a can cutter
7. Record the head space
8. Put all the contents over a sieve and let the liquid drain
9. Record the weight of empty can (B) to calculate net contents
10. Record the drained weight of the pieces.
11. Record whether the liquid is syrup or brine
12. Record the TSS of the salometer reading of the liquid (as the case may be)

Observations

Parameter	Observations
Name of the product	
Weight of can (A)	
Date of packing	
Expiry date	
Whether the can is bulged?	
Whether the can is leaking?	
Whether the lid of can is concave or convex?	
Vacuum in the can	
Head space	
Type of liquid (syrup or brine)	
Weight of empty can (B)	
Net weight of contents C = A - B	
Weight of the liquid (D)	
Drained weight (E)	
TSS of the liquid portion	
Acidity of the liquid portion	

For Teacher

a. Teacher shall tell the students the FPO specifications for various canned products

b. He/ she shall also explain them the stepwise process followed for the cut out analysis

c. He/she shall also ask the students to record the observation in tabulated form.

RELEVANT INFORMATION

The canning of fruits warrant a product of high quality, but the quality would depend on a number of internal and external attributes of a canned product. The can should not be puffed or bulged as this would indicate an undergoing fermentation reaction inside the can. Any pilferage or leakage of the contents also reduces the quality and makes it susceptible for spoilage.

The concavity of the lid would indicate a properly sealed and processed can and negate any incidence of fermentation going on inside the can. This would also indicate that there is a negative pressure i.e. vacuum inside the can which pulls the lid inside and gives a concave appearance to the can.

Fruits are generally canned in sugar syrup and vegetables in brine. The drained weight should meet the specifications of the canned products as laid down in FPO.

Suggested Readings

FPO 1991. The Fruit Products Order 1955. Ministry of Food Processing Industries, Government of India, 49p.

Ranganna S 1997. Handbook of Analysis and Quality Control for Fruit and Vegetable Products. 2nd edn. Tata McGraw-Hill Pub. Co. Ltd., New Delhi 1101p.

❑❑❑

Exercise 16

Dehydration of Fruits and Vegetables

LEARNING OBJECTIVES

The students should be able to

- Understand the principles of preservation used for drying and dehydration
- Understand various drying methods used for fruits and vegetables
- Dry and store the commodity for later use in various academic activities

Materials Required

Dehydrator, balance, thermometer, sulphur dioxide, blanching pan, fruits / vegetables, polythene or aluminium bags.

Procedure

1. Select sound fruits or vegetables
2. Record their initial weight and wash them properly in plain water to remove any dirt or extraneous matter.
3. Peel with stainless steel knife (as the case may be) and cut into thin uniform pieces or slices of 0.2-0.5 mm thickness
4. Blanch the slices in boiling water for 1-2 min. to inactivate the enzymes followed by immediate cooling.
5. Treat the slices with sulphur (sulphuring or sulphiting) in order to prevent darkening of the product during drying.
6. Dry the slices in a mechanical air dehydrator at about $55 \pm 5^{o}C$ temperature.
7. Record the weight of the product after every 2 hr till it becomes (almost) constant i.e. when there no more (significant) loss of moisture from the product.
8. Pack the product in ploythene or aluminium laminated pouches and seal them air tight.
9. If it is to be converted into powder, the dried product can be ground in a grinder before packing.
10. Label the packages and store them in a cool dry place.
11. Plot the drying rate curve of the product on a graph paper and analyse it.
12. Record the initial and final moisture content of the product (as discussed in Exercise No. 19).

Observations

Parameter	Observations
Name of fruit or vegetable	
Weight of fruits or vegetables taken	
Peeling performed or not (if yes, how?)	
Size of pieces	
Time taken for blanching	
Weight after blanching	
Tray load (Quantity of sample kept for drying in each tray)	
Weight at 2 hr interval (initial, 2hr, 4 hr, 6 hr, 8 hr ....)	
Total drying time taken	
Drying ratio	

For Teacher

a. The teacher shall tell the students the principles of preservation used for preparation of dried products.

b. He/she shall also explain them the stepwise process followed for the preparation of dehydrated fruits or vegetables

c. He/she shall also explain how to plot the drying curve and compute dehydration ratio.

RELEVANT INFORMATION

Drying/ Dehydration means the removal of water

Drying- Removal of water under the influence of non conventional energy sources like Sun or wind.

Dehydration- Process of removal of moisture by the application of artificial heat under controlled conditions of temperature, humidity and air flow.

Principle: Removal of moisture form a product reduces the water activity, under which the microorganisms can not survive and reproduce

Various equipments that are used for dehydration are mechanical drier, cabinet drier, vacuum drier (drying done under reduced pressures for faster drying and better quality), spray drier (used for drying juices and pulps), freeze drier etc.

Factors Affecting the Rate of Drying

1. Composition of raw material
2. Size, shape, arrangement and stacking of produce
3. Temperature, humidity and velocity of air
4. Pressure (barometric or under vacuum)
5. Heat transfer to surface (conductive, convective or radiative)

Sulphuring : The whole fruits, slices or pieces are exposed to the fumes of burning sulphur inside a closed chamber known as Sulphur Box for 30-60 min or in small air tight rooms.

The sulphur dioxide fumes act as disinfectant and prevent the oxidation and darkening of fruits on exposure and thus improve their colour. The fumes also act as a preservative, check the growth of molds etc. and prevent the cut pieces from fermenting while drying in Sun.

Sulphitation : Dipping of fruits / vegetables pieces / slices in sulphur solution before drying /dehydration.

Sweating : The process of staking dried fruits in boxes/ bins to equalize the moisture contents before final packaging.

Problems with the dried products: Most of the dehydrated products are highly hygroscopic in nature i.e. they have the tendency to absorb moisture from the surrounding environment during storage resulting in quality loss and reduced shelf life. Therefore, the dried products should be packed properly preferably in air tight containers and stored in a cool and dry place.

Suggested Readings

Fellows P 1988. Food Processing Technology- Principles and Practices. Ellis Horwood Ltd., Chichester, England. 505p.

❑❑❑

Exercise 17

Preparation of Fruit Wine (Cider)

LEARNING OBJECTIVES

The students should be able to

- Understand the chemical reactions taking place during alcoholic fermentation.
- Prepare wine from apple or any local fruit.
- Understand principle of preservation involved for preservation of wines.

Materials Required

Apple (or any local fruit), fruit mill, hydraulic press, weighing balance, Refractometer/sugar pectinol, diammonium hydrogen phosphate, *Saccharomyces cerevisiae* culture, Potassium metabisulphite (KMS) etc.

Procedure

1. Take a known quantity of apple (or any other fruit) say 20 Kg

2. Extract the juice as discussed in earlier exercises.
3. Add about 200 ppm of KMS.
4. Add sugar syrup to increase the TSS to about 20 – 24° Brix.
5. To about 10 Kg of juice diammonium hydrogen phosphate (10 g) and pectinol (20-25 g are added).
6. Take about 500 ml of this mixture in a conical flask and add pure *Saccharomyces cerevisiae* culture @ 1 % and plug the neck with cotton. Keep at 22-24° C to complete the fermentation within one day and yeast culture is ready.
7. Add this culture to about 10 Kg of must for further fermentation (i.e. culture is added @ 5% therefore 500 g is added to 10 Kg must).
8. Record the TSS of the must before fermentation.
9. Keep must for fermentation in big jars (narrow mouth) filled to about 75 % of their volume. Close the mouth of jars or bottles with cotton plug.
10. Allow fermentation to take place at 22-24° C for about 10–15 days.
11. When no more bubbles are coming out of the jars, fermentation is complete. At this stage the TSS reaches about 7° Brix.
12. Siphon the wine. Add filter aids if wine is not clear.
13. Blend the wine with sweetening agents if required.
14. Fill in beer bottles by keeping a head space of about 2.5 cm followed by crown corking them.
15. Pasteurize the bottles by keeping at 62.5° C for 20 min.

Observations

Parameter	Observations
Name of fruit	
Weight of juice taken	
Quantity and TSS of sugar syrup added	
Final adjusted TSS of the must	
Quantity of yeast culture added for fermentation	
Temperature of fermentation	
Final TSS of the wine when the fermentation is complete.	
Quantity of wine obtained after siphoning	
Temperature of pasteurization	
Duration of pasteurization	

For Teacher

a. The teacher shall tell the students the stepwise process followed for the preparation of fruit wines

b. He/she shall also tell the principles of preservation used in wines

RELEVANT INFORMATION

Wine: Product made by alcoholic fermentation of grapes or grape juice unless otherwise specified, by yeast and a subsequent ageing process. Alcohol content is 11-14 %, but may be as low as 7 %.

Alcoholic Fermentation

$$C_6H_{12}O_6 \xrightarrow[\text{Cal}]{\textit{Saccharomyces cerevisiae}} 2\,C_2H_5OH + 2\,CO_2 + 55\text{ K Cal}$$

Glucose Ethyl Alcohol

Principle of preservation: Normally wines do not spoil for years together as the carbohydrates (sugars) present in the juice have already been converted into alcohol and they are not available for the growth of spoilage causing microorganisms. Presence of alcohol is not suitable for growth of microorganisms and may cause desiccation and death of microorganisms.

Types of Wines

Still Wines: Retain no CO_2 eg cider etc

Sparkling Wines: Contain considerable amount of CO_2 eg. Champagne

Dry wines: Contain no or little unfermented sugar

Sweet wines: Contain unfermented sugars or is added later on.

Fortified Wines: Contain added alcohol / distillate of wine (brandy). Alcohol % is 19-21 %.

Table wines: low alcohol and little or no sugar

Dessert wines : Fortified sweet wines.

Sherry	Produced by special technique (secondary fermentation by special type of "Flour yeast" or by baking), containing 18-21% alcohol, could be sweet or dry.	
Cider	Apple wine.	
	Soft Cider	1-5 % alcohol
	Hard Cider	5-8 % alcohol
	Apple wine	> 8% alcohol but can be as high as 14%.
Perry	Pear wine (could be sweet or dry)	
Mead	Honey wine	

Boukha	Fig wine
Toddy	Coconut wine
Vodka	Prepared from potato
Fanny	Cashew apple wine
Port wines	Made from grape must fortified with brandy. Produced in Douro region of Portugal (alcohol about 18%).
Vermouth	Fortified wine with 15-21 % alcohol, flavoured with a characteristic mixture of herbs and spices, some of which impart an aromatic flavour and odour while others a bitter flavour (may be sweet or dry).
Brandy	It is an alcoholic distillate from fermented grape juice or wine unless otherwise specified.
Rum	Distillate from alcoholically fermented sugarcane juice or molasses
Whiskey	Distillate from fermented grains, rye, wheat.
Beer	Prepared from Barley grains

Precautions

1. Always select sound fruits free from any kinds of diseases, rotting or spoilage etc for the extraction of juice.
2. If the wine is being prepared from stored juice, make sure that the juice is sound and fit for consumption.
3. Always use properly washed and dried stainless steel utensils for cooking heating etc.
4. Always use pre-sterilized glass bottles or jars for fermentation.

5. Plug the mouth of the bottles and jars quickly with cotton.
6. Pasteurize the wine properly before storage.

Suggested Readings

Joshi VK, Chauhan SK and Bhushan S 2000. Technology for fruit based alcoholic beverages. *In:* Postharvest Technology of Fruits and Vegetables - Handling, Processing, Fermentation and Waste Management. Verma LR, Joshi VK (eds.) Indus Pub. Co. New Delhi, pp 1019-1101.

Joshi VK 1998. Fruit Wines. Directorate of Extn. Edu., Dr Y S Parmar Univ. Hort. Fty., Nauniu, Solan, HP, India, 226p.

Exercise 18

To Study Extension of Vase Life of Flowers

LEARNING OBJECTIVES

The students should be able to

- ❒ Understand the principles of extension of shelf life of cut flowers in vase solutions.
- ❒ Understand various chemicals and their concentrations used for preparing vase solutions.
- ❒ Lay out an experiment for studying the vase life of different cut flowers.

Materials Required

Cut flowers, 1000 ml conical flasks, scissor, tape, distilled water, chemicals, weighing balance, volumetric flasks, labels, thermometer (max. & min. type) and hygrometer.

Procedure

1. Collect fresh spikes of flowers
2. Cut at the base keeping about 25-60 cm of the cut flowers stick (uniformly for each flower).

3. Prepare Pulsing/ Vase solutions.
4. Put the prepared flower spikes in labeled conical flasks containing different Pulsing/ Vase solutions (for a known period of time).
5. Record the weight of the spikes and other observations daily.
6. Record the temperature and relative humidity of the storage chamber daily.
7. Analyse the recorded data and draw conclusion about the best vase solutions for the selected cut flower.

Observations

Name of flower
Length of spike retained for experiment
Volume of vase solutions being used

Parameter	**Pulsing/Vase solutions**										
	1	**2**	**3**	**4**	**5**	**6**	**7**	**8**	**9**	**10**	**...**
Initial Weight (g)											
Solution uptake (ml)											
Days to open I floret											
Days to wither I floret											
Total No of opened florets											
Total No. of withered florets											
Days to wither all florets											
Final weight											
Rotting incidence (if any)											

For Teacher

a. The teacher shall tell the students the difference between pulsing and vase solutions and principles of extension of shelf life in such solutions.

b. He/she shall also explain them the stepwise process to be followed for the experiment and recording observations.

RELEVANT INFORMATION

Cut flowers: Flower which is cut alongwith a portion of the stem. The longevity of cut flowers depends on carbohydrate reserves of flowers, osmotic concentration and pressure potential of petal cells, stomatal functioning, differences in the numbers of thick walled cells in the xylem and phloem, presence of various hormones and susceptibility to diseases and insect pests.

Harvesting of flowers should be done at optimum stage as follows

Name of flower	Stage of harvesting
Gladiolus	1-5 florets show colour
Rose	1-2 petals begin to unfold
Lily	Coloured buds
Chrysanthemum (standard type)	Outer florets fully expanded
Carnation (standard type)	Paint brush stage
Tulip	Half coloured buds

Pulsing solutions: Pulsing refers to the short duration pre-shipment or pre-storage treatment given to the flowers generally for about 16-24 hours to extend the longevity of the flower. Sucrose (sugar) in high concentrations (10-30%) is the important component of pulsing solutions.

Holding / Vase solutions: Vase solutions are made to hold flowers continuously till the termination of their vase life or senescence. The concentrations of sucrose in vase solutions is kept low (0.5-2.0 per cent).

Floral preservatives: Any chemical formulation which is used for extending vase life of flowers. The main constituents of floral preservatives are as follows:

a. *Water*: Tap water may be harmful due to its alkalinity, presence of dissolved solutes and toxic ions. So distilled water or boiled and decanted water should be used for preparing vase/pulsing solutions.

b. *Sugar*: It acts as additional food source and improves water balance of cut flowers. For pulsing solutions sugar @ 10-30 % can be used while that for vase or holding solutions it should be used @ 0.5 to 2.0 %. Since it promotes microbial growth, it may be combined with biocides

c. *Biocides*: 8-hydroxyquinoline citrate (8-HQC) @ 200-600 ppm, $AgNO_3$ @ 25 ppm (although this is considered to be hazardous to environment), Aluminium sulphate @ 100-300 ppm, Citric acid @ 50-1000 ppm can be used as biocides.

Suggested Readings

Kushal S, Arora JS and Bhattacharjee SK 2001. Postharvest Management of Cut Flowers. Technical Bulletin No. 10, AICRP on Floriculture, Div., of Floriculture and Landscaping, IARI, New Delhi p39.

❑❑❑

Exercise 19

Determination of Moisture and Total Solids

LEARNING OBJECTIVES

The students should be able to

- ❐ Understand the principle used for determination of moisture and total solids
- ❐ Learn the stepwise procedure used for determination of moisture and computation of total solids.

Materials Required

Fruit, vegetable or product sample, hot air oven, weighing balance, crucibles, petri plates, lead pencil / marker etc.

Procedure

1. Take crucibles or petri plates and record the tare weight of crucibles / petri plates (C) and mark them

2. Cut the sample into small and thin pieces (if solid)
3. Put the sample into the crucibles / petri plates in three replications
4. Record the weight of crucibles + sample (C + S)
5. Put in hot air oven for drying at 70 ± 2 °C.
6. Record the loss in weight after every 1-2 hrs till it becomes constant
7. Calculate the moisture and total solids.

Observations

- weight of individual crucibles used (denote as C_1, C_2, C_3, C_4, ...)
- weight of crucible + sample (denote as C_1+S_1, C_2+S_2, C_3+S_3, C_4+S_4, ...)

Time (initial)	C_1+S_1	C_2+S_2	C_3+S_3	C_4+S_3	C_5+S_5	C_6+S_6	C_7+S_7
1 hr							
2hr							
3hr							
-							
-							
Final (when weight becomes constant)							

For Teacher

a. The teacher shall tell the students the difference between moisture and total solids and the stepwise procedure for their estimation

b. He/ she shall also tell the students the principle used for determination of moisture and total solids.

c. The teacher shall also explain the conditions under which the moisture or total solid values exceed 100.

RELEVANT INFORMATION

Principle: Heating of sample evaporates the water from it. The weight loss due to evaporation is moisture and the weight left over is the total solids.

Moisture is the total quantity of water present in the sample. Total solids is the weight of sample deducting the moisture. It is measured by drying the sample at 70 $\pm$ 2 °C temperature in oven. The reduction in weight is computed as moisture %.

Sample weight = Moisture + Total Solids

Moisture (%) = 100 – Total Solids (%)

Total Solids (%) = 100 – Moisture (%)

Calculations

Let C_1 be 50 g, C_1+S_1 (initial) be 70 g and $C+S_1$ (final) be 54 g

$$\text{Moisture \% (fresh weight basis)} = \frac{\text{(Initial weight - Final weight)}}{\text{Initial weight}} \times 100$$

$$= \frac{((70\text{-}50) - (54\text{-}50)) \times 100}{(70\text{-}50)}$$

$$= \frac{(20\text{-}4)}{20} \times 100 = 80\%$$

$$\text{Moisture \% (dry weight basis)} = \frac{\text{(Initial weight - Final weight)}}{\text{Final weight}} \times 100$$

$$= \frac{((70\text{-}50) - (54\text{-}50)) \times 100}{(54\text{-}50)}$$

$$= \frac{(20\text{-}4)}{4} \times 100 = 400\ \%$$

Note: Moisture on fresh weight basis can never be > 100 but on dry weight basis it may exceed 100 as in the cited example.

$$\text{Total solids \%} = \frac{\text{Final weight}}{\text{Initial weight}} \times 100$$

$$= \frac{(54\text{-}50))}{(70\text{-}50)} \times 100 = 20\ \%$$

$$\text{or Total Solids \%} = 100 - \text{moisture \%}$$

$$= 100 - 80 = 20\%$$

❏❏❏

Exercise 20

Determination of Minerals as Total Ash

LEARNING OBJECTIVES

The students should be able to

- Understand the principle used for determination of minerals as total ash
- Learn the stepwise procedure used for determination of ash

Materials Required

Sample, hot air oven, weighing balance, crucibles, muffle furnace, lead pencil / marker etc.

Procedure

1. After following the procedure for the determination of moisture (in silica crucibles as given in previous chapter) the same sample can be used for the determination of ash.
2. Heat the dried sample in the silica crucibles over a heater/hot plate/burner to burn the carbon.

3. When it becomes completely black and no more smoke comes out of the sample, shift them to muffle furnace.
4. After heating the samples in muffle furnace, all the marks of pen / pencil/ marker are lost, so one would not be able to identify the samples and error of weights may occur. Therefore, it is advised to place the samples in the furnace in one or two rows and draw the diagram on a piece of paper for identification purpose.

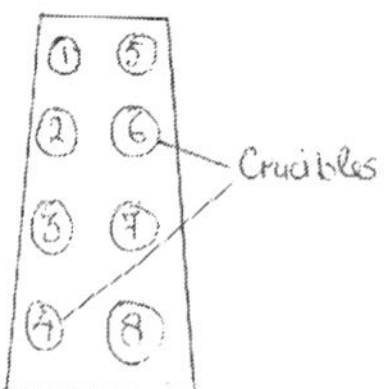

5. Turn the furnace on and heat the samples at about 500-700 °C for 5-6 hrs or more if required.
6. Cool the furnace by keeping it off overnight.
7. Record the weights of crucibles individually.
8. The difference in weights gives the total ash content and expressed in %.

Observations

- weight of individual crucibles used (denote as C_1, C_2, C_3, C_4, ...)
- weight of crucible + sample (denote as C_1+S_1, C_2+S_2, C_3+S_3, C_4+S_4, ...)

Time	C_1+S_1	C_2+S_2	C_3+S_3	C_4+S_4	C_5+S_5	C_6+S_6
Fresh weight of sample						
Weight after drying in oven						
Weight after burning in muffle furnace						

For Teacher

a. The teacher shall tell the students the principle used for estimation of minerals as ash

b. He/ she shall also tell the students the stepwise procedure for estimation of ash.

RELEVANT INFORMATION

Principle: On heating plant tissues to very high temperatures, the water (H_2O) is evaporated and carbon (C) is burnt. The left over white powdery mass is basically the minerals which are not damaged even after heating to very high temperatures and is called ash. This gives the estimate of minerals present in the plant tissue.

$$\text{Total Ash \% (Fresh weight basis)} = \frac{\text{Ash weight}}{\text{Initial weight}} \times 100$$

$$\text{Total Ash \% (Dry weight basis)} = \frac{\text{Ash weight}}{\text{Dry weight}} \times 100$$

Let the weight of ash be 0.25g

$$\text{Total Ash \% (Fresh weight basis)} = \frac{0.25}{20} \times 100 = 1.25\%$$

$$\text{Total Ash \% (Dry weight basis)} = \frac{0.25}{4} \times 100 = 6.25\%$$

Suggested Readings

Ranganna S 1997. Handbook of Analysis and Quality Control for Fruit and Vegetable Products. 2nd edn. Tata McGraw-Hill Pub. Co. Ltd., New Delhi 1101p.

❐❐❐

Exercise 21

Determination of Total Soluble Solids (TSS)

LEARNING OBJECTIVES

The students should be able to

- Understand the difference between total solids and total soluble solids
- Learn the procedure used for determination of TSS
- Learn the effect of sample temperature on TSS

Materials Required

Fruit/ sample, refractometers, thermometers, standard solutions of known TSS, temperature correction Tables.

Procedure

1. Calibrate refractometers of all ranges to be used.
2. The TSS of the juices/pulps/fresh fruits can be measured with the one having lowest range i.e. 0-32°

Brix while, for measuring the TSS of concentrates, syrups, squashes, jams, jellies etc. refractometers of the higher ranges are made use of.

3. Put a drop of material on to the glass surface of refrectrometer's prism, cover gently with the lid and observe the TSS against light.
4. Measure the temperature of the liquid which is being evaluated and apply temperature corrections (Annexure III).
5. Record the corrected value as the TSS of the sample.

Observations

	Sample No.				
	1	**2**	**3**	**4**	**5**
Temperature of sample					
Original measured TSS value					
TSS value after applying temperature corrections					

For Teacher

The teacher shall tell the students

a. The difference between TSS and total solids
b. Effect of temperature on TSS values
c. Methods of calibration of refractrometer.

RELEVANT INFORMATION

The TSS of various fruits/ juices/ pulps are generally measured by Hand refractometer and expressed in °Brix or percentage. There are generally different ranges of refractometers available in the market i.e. 0-45 and 45-85° Brix, and 0-32, 28-62, 58-92° Brix. In any of the cases, the refractometer of the lowest range i.e. 0-45 or 0-32 is to be calibrated at 0° with distilled water at 20° C. For calibrating the higher ranges of refractometers standard solutions having known ^{o}Brix are to be prepared and the refractometers are calibrated accordingly at 45 or 28 or 58° Brix as applicable.

Total solids = Total soluble solids + total insoluble solids

- Generally, the acids, sugars, vitamins, amino acids and other soluble solids present in the fruits or product contribute partially to the TSS content.
- The values of TSS as recorded by hand refractometers decrease with the increase in temperature.
- If a sample has TSS of 15° Brix at 20° C temperature, the same sample shall show a TSS of 14.22^{o} Brix at 30° C and 15.61^{o} Brix at 10° C.
- The prism of the refractrometer should always be properly washed and dried before recording observations.
- Very hot samples (80-90° C), may not give clear cut value due to formation of steam on the prism of the instrument.
- TSS (%) can never be more than the total solids (%) of the same sample.

❑❑❑

Exercise 22

Determination of Titratable Acidity

LEARNING OBJECTIVES

The students should be able to

- Understand the principle used for estimation of titratable acidity
- Learn the stepwise procedure used for determination of titratable acidity
- Learn the relationship between TSS and acidity of the sample

Materials Required

Sample, 0.1 N NaOH solution, 1% phenolphthalein indicator solution, beakers, conical flasks, pestle and mortar, funnels, volumetric flasks, pipettes, pH meter, shaker.

Reagents

1. *0.1 N NaOH:* Dissolve 4 g NaOH (AR) crystals in sufficient water to make up the final vol. 1 litre.

2. *1 % Phenolphthalein:* Dissolve 1 g in 100 ml water. Readymade indicator solutions are more commonly used.

Procedure

Preparation of sample: For fresh fruits and vegetables, prepare a well blended, uniform pulp in a pestle and mortar and for processed products, take out a representative sample and blend it well so that it becomes uniformly soluble when added to water. Filter through a fine muslin cloth or coarse filter paper if felt necessary. The liquid samples can be taken directly for estimation.

1. Take out a known weight / vol. (say 20 g) of sample in a beaker (in 3 replicates).
2. Make it to a known vol. (say 100 ml) with added distilled water.
3. Take 10 ml of this aliquot in a conical flask.
4. Add 1-2 drops of phenolphthalein indicator into the aliquot.
5. Titrate against 0.1 N solution of NaOH till light pink coloured end point is obtained.
6. For coloured products, i.e. tomato products where judgment of light pink colour is difficult, titration may be done over a shaker, and digital pH meter may be used for continuously measuring pH of aliquot which would show a steady increase with more and more addition of 0.1 N NaOH. Titre value may be recorded when pH of 8.3 is obtained.
7. Record the titre value and calculate the acidity of the sample as per the given formula.

$$\text{Titratable Acidity \%} = \frac{\text{Titre} \times \text{Normality of alkali} \times \text{Volume made up} \times \text{Equivalent weight of acid} \times 100}{\text{Wt./Vol. of sample (W)} \times \text{Vol. of aliquot} \times 1000}$$

Observations

	Sample No.				
	1	2	3	4	5
TSS of the sample					
Weight of sample					
Vol. made					
Vol. of aliquot taken for titration					
Titre value					
Equivalent weight of acid					
Acidity as calculated					
Whether the TSS > acidity or not					

For Teacher

The teacher shall tell the students

a. The principle underlying the estimation of titratable acidity

b. How the acidity contributes in the TSS of a sample

c. Stepwise procedure followed for the estimation of acidity.

RELEVANT INFORMATION

Principle: The acidity of solution is neutralized by addition of alkali i.e. NaOH solution. Therefore the pH of the sample aliquot increases continuously while addition of alkali. The phenolphthalein solution shows pink colour at pH 8.3-10.0. As the pH of the sample aliquot reaches 8.3

the indicator shows pink colour. This pink colour can be reverted back to its original, on further addition of acid in the sample.

The Equivalent weight of different acids and the predominant acids present in various fruits:

Fruit/Vegetable/ products	Predominant acid	Equivalent weight
Apple, Pear, peach, plum, apricot, banana, cherry	Malic acid	67
Citrus, lemon, orange, lime, currants, fig, gooseberry, guava, pineapple, pomegranate, strawberry, raspberry, mango	Citric acid	64
Grapes	Tartaric acid	75
Pickles	Acetic acid	60

- Malic acid (HOOC-CH_2-CHOH-COOH), Mol. Wt. 134.09, Valency 2, Equivalent wt. 67.05
- Citric acid (HOOC-CH_2-C(OH)(COOH)-CH_2-COOH), Mol. Wt. 192.12, Valency 3, Equivalent wt. 64.04
- Tartaric acid (HOOC-CHOH-CHOH-COOH), Mol. Wt. 150.08, Valency 2, Equivalent wt. 75.04
- Acetic acid (CH_3-COOH), Mol. Wt. 60.05, Valency 1, Equivalent wt. 60.05
- Lactic acid (CH_3-CHOH-COOH), Mol. Wt. 90.08, Valency 1, Equivalent wt. 90.08.

Suggested Readings

Hulme, A C 1971. The Biochemistry of Fruits and Their Products. Vol 1 & 2, Acad. Press London.

Ranganna S 1997. Handbook of Analysis and Quality Control for Fruit and Vegetable Products. 2nd edn. Tata McGraw-Hill Pub. Co. Ltd., New Delhi pp. 9-10.

Gould, W A 1977. Food Quality Assurance. The AVI Pub Co., Westport Connecticut. 314p.

❑❑❑

Exercise 23

Determination of Reducing and Total Sugars by Nelson - Somogyi Method

LEARNING OBJECTIVES

The students should be able to

- Differentiate between reducing, non-reducing and total sugars
- Learn the stepwise procedure used for determination of sugars
- Learn the relationship between TSS and sugar content of the sample.

Materials Required

Sample, reagents, beakers, conical flasks, pestle and mortar, funnels, volumetric flasks, pipettes, hot plate, weighing balance, spectrophotometer/colorimeter.

Reagents

1. 0.1 N NaOH: Dissolve 4 g NaOH (AR) crystals in sufficient water to make up the final volume 1 Litre.
2. 1 % Phenolphthalein: Dissolve 1 g in 100 ml water. Readymade indicator solutions are more commonly used.
3. Alkaline copper Tartarate
   a. Dissolve 12.5 g anhydrous sodium carbonate, 10 g sodium bicarbonate, 12.5 g potassium sodium tartarate and 100 g anhydrous sodium sulphate in about 400 ml of distilled water and make up the final volume to 500 ml.
   b. Dissolve 15 g copper sulphate in distilled water. Add one drop of sulphuric acid and make up to 100 ml in distilled water.
4. Always prepare fresh Alkaline Copper Tartarate reagent by mixing 96 ml of sol. a. and 4 ml of sol. b. and before use.
5. Arsenomolybdate Reagent: Dissolve 5.0 g ammonium molybdate (AR) in 90 ml of distilled water. Add about 5.0 ml of concentrated sulphuric acid and mix well. Add 600 mg of disodium hydrogen arsenate dissolved in 50 ml water. Mix well and incubate the prepared reagent at 37°C for 24-48 hr or 55° C for 30 min. with constant stirring.

Procedure

1. Weigh about 100-200 mg of the sample and extract the sugars with hot 80% ethanol twice or thrice (about 5 ml each time).
2. Collect and pool the supernatants and evaporate by keeping it on a hot water bath at 80±2° C.

3. Add 10 ml water and dissolve the sugars over a Cyclomixer.
4. Pipette out aliquots of 0.1, 0.2 and 0.5 ml into separate test tubes.
5. Make up the volume to 2 ml with distilled water.
6. Pipette out 2 ml distilled water as blank.
7. Add 1 ml Alkaline Copper Tartarate reagent to the blank and the sample tubes.
8. Boil the tubes on a water bath for 10 min.
9. Cool the tubes under running water and add 1 ml arsenomolybdate reagent to all the tubes.
10. Make up the volume to 10 ml with distilled water and keep for 10 min.
11. Read the absorbance of blue colour at 620 nm.
12. Calculate the amount of sugars present by making use of the formula given below.

Note: If the sugar concentration of the sample is too high the dilution at step 3 may be increased and the same may be considered while making calculations.

Preparation of Standard Curve

1. Pipette out 0.2, 0.4, 0.6, 0.8 and 1 ml of the working standard in a series of test tubes.
2. Follow steps 5 to 11 given above.
3. Plot the values of the absorbance against different concentrations of glucose.
4. Draw the line of best fit.
5. Compute the value of sugars for unit absorbance (say Z).

% reducing sugars = absorbance of sample x Z x Dilution factor

Total sugars: For estimation of total sugars, take about 25 ml of sample and keep it overnight after adding about 5-10 ml of HCl (1:1) solution. Invert the sugars by boiling it the next morning, neutralize the acid by using IN NaOH and make up the volume to 50 or 100 ml. Estimate the sugars of the sample as in case of reducing sugars. Apply the dilution factor while making calculations.

Observations

Parameter	Sample No.					
	1	2	3	4	5	6
Sugar concentration for unit absorbance						
Dilution factor						
Absorbance of sample						
Reducing sugar (%)						
Total sugar (%)						
Non – reducing sugar (%)						

For Teacher

The teacher shall tell the students

a. The principle underlying the estimation of sugars
b. How sugars contribute in the TSS of the sample
c. How to handle spectrophotometer and the precautions to be taken
d. Stepwise procedure to followed for the estimation of sugars

RELEVANT INFORMATION

Principle: The reducing sugars i.e. glucose, fructose when heated with alkaline copper tartarate reduce the copper from the cupric to cuprous state, thus forming cuprous oxide. When cuprous oxide is treated with arsenomolybdic acid, the molybdic acid reduces to to molybdenum (blue). The blue colour developed is compared with a set of standards in a colorimeter at 620 nm. Non-reducing sugars i.e. sucrose when heated in the presence of acid are converted into reducing sugars i.e. glucose and fructose. Therefore, the non reducing sugars can also be estimated as reducing sugars after their inversion. Total sugars are estimated as reducing sugars after inversion of all the sugars i.e. non-reducing sugars into reducing sugars and thus their estimation.

Suggested Readings

Sadasivam, S. and Manickam, A. 1996. Biochemical Methods, 2nd edn. New Age International (P) Ltd. Pub. New Delhi pp5-6.

Ranganna, S. 1997. Handbook of Analysis and Quality Control of Fruit and Vegetable Products. 2nd Edn. Tata McGraw Hill Publishing Co., New Delhi. pp. 12-19.

❑❑❑

Exercise 24

Determination of Ascorbic Acid

LEARNING OBJECTIVES

The students should be able to

- ❒ Determine the ascorbic acid contents of various fresh and processed samples
- ❒ Learn how to remove interference of SO_2 in the samples for correct estimation of vitamin C.

Materials Required

Sample, reagents, beakers, conical flasks, pestle and mortar, funnels, volumetric flasks, pipettes, weighing balance, spectrophotometer/colorimeter.

Reagents

1. 2% Metaphosphoric Acid Solution: Weigh 20 g of HPO_3 and dilute to 1 litre in water or 4% oxalic acid.
2. Dye Solution: Dissolve 100 mg of 2, 6 Dichlorophenol-indophenol dye (Phenolindo-2,6- dichlorophenol

sodium salt, AR, $C_{12}H_6Cl_2NNaO_2$ 2-3H_2O) and 84 mg of sodium bicarbonate (sodium hydrogen carbonate AR, $NaHCO_3$) in hot (85-95°C) distilled water. Cool and make up the vol 100 ml. Filter and dilute 25 ml of the dye to 500 ml with distilled water.

3. Standard Ascorbic Acid: Weigh 50 mg of ascorbic acid (AR) and make up to 50 ml with 2% metaphosphoric acid or 4% Oxalic acid solution (1000 ppm i.e. 1000 µg/ml). Dilute 4 ml of this solution to 100 ml with 2% HPO_3 or 4% oxalic acid. (1ml = 40 µg ascorbic acid).

Procedure

1. Take 10-20g of sample (A), grind and make up to 100 ml (B) with 2% HPO_3 or 4% oxalic acid.
2. Filter through filter paper
3. Take out 2-3 ml (C) of aliquot and make 5 ml with 2% HPO_3 or 4% oxalic acid in test tubes.
4. Prepare a blank with 5 ml of 2% HPO_3 or 4% oxalic acid and 10 ml water.
5. Add 10 ml of dye solution to all the sample tubes and shake immediately in a shaker.
6. Record the absorbance of red colour (pink) at 518 nm within 15-20 seconds
7. Plot the standard curve and calculate the ascorbic acid concentration (X) of the sample from it.

Standard Curve

1. Pipette 1.0, 1.5, 2.0, 2.5, 3.0, 3.5, 4.0, 4.5 and 5.0 ml of standard ascorbic acid solution (1ml = 40 µg ascorbic acid) to a series of test tubes.

2. Make up to 5 ml with 2% HPO_3 or 4% oxalic acid.

3. Follow the steps 4 to 6 above.
4. Record the absorbance for each concentration and plot against each other.
5. Draw the line of best fit.
6. Calculate the ascorbic acid concentration for a unit absorbance value.

$$\text{mg ascorbic acid per 100 ml} = \frac{\text{Ascorbic acid Content (X)} \times \text{Volume made up (B)} \times 100}{\text{Wt/ Vol of sample (A)} \times \text{Vol of solution taken for estimation (C)} \times 1000}$$

Observations

Parameter	Sample No.				
	1	**2**	**3**	**4**	**5**
Absorbance of sample (Z)					
Ascorbic acid content (X)					
Weight / Vol of sample (A)					
Volume made up (B)					
Vol of solution taken for estimation (C)					
Ascorbic acid content (mg / 100 g/ml)					

For Teacher

The teacher shall tell the students

a. The principle underlying the estimation of ascorbic acid
b. How to handle spectrophotometer and the precautions to be taken
c. Stepwise procedure to followed for the estimation of ascorbic acid

RELEVANT INFORMATION

Principle: This method is based the measurement of the extent to which a 2,6 dichlorophenol indophenol sodium salt solution is decolourised by ascorbic acid in sample extracts and standard solutions. Since interfering substances reduce the dye slowly, rapid determination would be by and large measuring the ascorbic acid.

Sulphur dioxide when present in the juice or food sample reduces the 2,6 dicholorphenol indophenol dye and interferes in ascorbic acid analysis. Eliminate the interference of SO_2 by addition of 1 ml of 40% formaldehyde and 0.1 ml HCl to about 10 ml of filtrate sample (B) and keeping for 10 min.

Note: Since the colour of the samples in test tubes changes continuously the observations should be recorded individually i.e. add 10 ml of dye (step 5) to I sample and record the absorbance and then add the dye to the II sample and record absorbance followed by III, IV and so on. Do not add dye to all samples at once and then record the absorbance values together as this may not give the exact readings. The time taken between addition of dye to the sample and recording the absorbance, both, during the preparation of standard curve and analysis of samples, should be the same for every sample as the readings of the same sample may change continuously even in the spectrophotometer.

Suggested Readings

Ranganna, S. 1997. Handbook of Analysis and Quality Control of Fruit and Vegetable Products. 2nd edn. Tata McGraw Hill Publishing Co., New Delhi. pp. 106-107.

❑❑❑

Exercise 25

Determination of Specific Gravity

LEARNING OBJECTIVES

The students should be able to

- ❒ Able to determine the specific gravity of liquid samples
- ❒ Understand how specific gravity changes with the temperature.

Materials Required

Specific gravity bottles, weighing balance, sample, pipettes, thermometer, distilled water

Procedure

1. Take the sample and distilled water record their temperatures (they should be same for correct estimation). If the temperatures are not same bring both to 20°C or the room temperature.

2. Clean and dry the specific gravity bottles properly before use.
3. Record the tare weight of the specific gravity bottle (T) alongwith the stopper.
4. Carefully fill the specific gravity bottle with distilled water up to the brim and then place the stopper tightly so as to exclude extra water.
5. Dry the exteriors of the bottle with a tissue paper
6. Record the weight of (bottle + water) say A
7. Similarly, fill the same specific gravity bottle with test sample and record the weight (bottle + sample) say B
8. Calculate the specific gravity of solution using the following formula

$$\text{Specific gravity} = \frac{B - T}{A - T}$$

9. For individual fruits like orange, first the weight of fruit in air (p) is noted. The same fruit is then put in a beaker (1-2 litre capacity) containing some quantity of water. The fruit is submerged completely by pushing and holding in water with the help of a wire loop. The weight of fruit in submerged condition (q) is also recorded. The ratio of the weight of fruit in air to its submerged weight would be its specific gravity and is generally less than 1.0.

$$\text{Specific gravity} = \frac{\text{Weight of fruit in air}}{\text{Weight of fruit in water}} = \frac{p}{q}$$

Observations

Parameter	Sample No.				
	1	2	3	4	5
Temperature of distilled water					
Temperature of sample solution					
Tare weight of bottle (T)					
Weight (Bottle + water) A					
Weight (bottle + sample) B					
Specific gravity					
Weight of fruit in air (p)					
Weight of fruit in water (q)					
Specific gravity of fruit					

For Teacher

The teacher shall tell the students

a. Steps for determining the specific gravity of liquid

b. How specific gravity changes with the increase or decrease in temperature

c. Why the temperatures of water and the sample should be the same for determination of specific gravity

RELEVANT INFORMATION

Specific gravity is the ratio of the density of a material to the density of water. Specific gravity has no units since it is a relative value. Density of water is 1 g/cm^3 or 1000 Kg /m^3.

Specific gravity of the fruits and vegetable juices containing lesser of alcohol or essential oils is always greater than unity. The specific gravity directly depends on the soluble solids. The greater the soluble solids the more is the specific gravity.

The specific gravity may also change with the increase in temperature as this may lead of the expansion of the liquids and the actual weight of the sample accommodated in the specific gravity bottle may vary.

The temperatures of distilled water and the sample liquid should always be the same for correct estimation. For this the samples can be held in water bath maintained at a specific temperature.

Suggested Readings

Ranganna, S. 1997. Handbook of Analysis and Quality Control of Fruit and Vegetable Products. 2nd edn. Tata McGraw Hill Publishing Co., New Delhi. 1112p.

Ting SV and Rouseff RL 1986. Citrus Fruits and Their Products. Marcel Dekker Inc., New York, 293p.

❑❑❑

Exercise 26

Determination of Relative Viscosity

LEARNING OBJECTIVES

The students should be able to

- Determine the relative viscosity of juices or liquid samples
- Understand how relative viscosity changes with the temperature

Materials Required

Ostwald viscometer, stop watch, sample, distilled water, thermometer.

Procedure

a. Take sample and distilled water. Record their temperatures (they should be same for correct estimation). If the temperatures are not same bring both to a specific temperature say 20°C or the room temperature.

b. Clean and dry the Ostwald viscometer properly before use

c. Fill the ostwald viscometer with distilled water and record the time taken (T_1) for the a known volume of water to flow through the orifice.

d. Empty the viscometer and dry it

e. Fill the ostwald viscometer with sample juice or liquid and record the time taken (T_2) for the a known volume of sample to flow through the orifice of the same viscometer.

f. Calculate the relative viscosity using the following formula

$$\text{Relative Viscosity} = \frac{T_2}{T_1}$$

Observations

Parameter	Sample No.				
	1	2	3	4	5
Temperature of distilled water					
Temperature of sample solution					
T_1					
T_2					
Relative Viscosity					

For Teacher

The teacher shall tell the students

a. Steps for determining the relative viscosity of liquids

b. How relative viscosity changes with the increase or decrease in temperature

c. Why the temperatures of water and the sample should be the same for determination of relative viscosity

❑❑❑

Exercise 27

Determination of Colour and Non-Enzymatic Browning

LEARNING OBJECTIVES

The students should be able to

- Determine the colour and non-enzymatic browining of samples

Materials Required

Spectrophotometer, ethyl alcohol, beakers, pestle and mortar etc.

Procedure

1. Preparation of sample extracts
   a. *Liquids* : Centrifuge the juice at 5000 rpm for 4-5 min., and collect the supernatant. For liquid samples prepare a 40 % solution in ethyl alcohol. To 30 ml of the sample add 45 ml of ethyl alcohol and mix thoroughly. Filter through Watman No. 1 paper.

   b. *Semi-solid samples:* To 15 g of blended sample add 15 ml of distilled water and 45 ml of ethyl alcohol. Mix thoroughly and filter through Watman No. 1 paper.

   c. *Dried products:* Extract known quantity of sample (4-6 g) with 100 ml of 60% ethyl alcohol for 12 hrs and filter through Watman No. 1 filter paper.

2. Measure the absorbance of the samples at 440 nm for non-enzymatic browning and 400 or 420 nm for measurement of colour.

3. Prepare 60% ethyl alcohol solution as blank. Add 30 ml of distilled water to 45 ml of absolute alcohol and use as blank.

Observations

Parameter	Sample No.				
	1	2	3	4	5
Wt / Vol. of sample taken					
Qty of water added if any					
Qty of alcohol added					
Whether filtration required or not?					
Absorbance at 440 nm (NEB)					
Absorbance at 420 nm (Colour)					

Arrange the samples in decreasing order of absorbance and conclude which sample is the lightest and darkest in colour. Does your observation match with your visual observation?

For Teacher

The teacher shall tell the students

a. Steps for preparation of sample extracts

b. How to record the absorbance for colour and non-enzymatic browning.

Suggested Reading

Ranganna, S. 1997. Handbook of Analysis and Quality Control of Fruit and Vegetable Products. 2nd Edn. Tata McGraw Hill Publishing Co., New Delhi. pp. 891.

❒❒❒

Exercise 28

Determination of Crude Fibre

LEARNING OBJECTIVES

The students should be able to

- ❑ Understand what is crude fibre
- ❑ Learn the stepwise procedure used for determination of crude fibre in plant samples

Materials Required

Sample, Liebig condenser, digestion flasks (700-750 ml cap), conical flasks, filtering cloth (thick linen cloth for clear filtration) reagents, muffle furnace, crucibles.

Reagents

1. 0.255 N Sulphuric acid solution : Dilute 2.5 g (1.36 ml) of concentrated H_2SO_4 to 200 ml. Ensure correct concentration by titration.

2. 0.313 N Sodium hydroxide solution: 2.504 g NaOH / 200 ml water nearly free from sodium carbonate. Ensure correct concentration by titration.
3. *Asbestos Chips:* gooch grade, medium fibre, acid wash and ignited
4. 10% Potassium sulphate (K_2SO_4) solution: Dissolve 10 g of Potassium Sulphate (AR) in water and make up to 100 ml.

Procedure

We follow 4 stages of washings i.e. with ether (Step 1), acid (Step 2-6), alkali (Step 7-11) and alcohol (Step 12) to obtain crude fibres along with some minerals. Crude fibre is burnt & mineral ash is obtained (Step 14-16). Crude fibre is estimated as weight loss after heating in muffle furnace.

1. Extract 2-3 g (X) of dry and powdered material with ether for removal of fats.
2. Transfer the residue and approx. 0.5 g asbestos to a digestion flask
3. Add 200 ml boiling sulphuric acid
4. Connect the digestion flask to the condenser and heat so that the solution comes to boiling within 1 min.
5. Rotate the flask frequently during brisk boiling for exactly 30 min.
6. Remove the flask after 30 min. and filter through linen in a fluted funnel. Wash with boiling water till the washings are no longer acidic.
7. Heat NaOH solution to boiling in a reflux condenser. Wash the residue from the acid digestion flask back into the flask with 200 ml of boiling NaOH. Connect the flask with reflux condenser and boil exactly for 30 min.

8. After boiling for 30 min. remove the flask and filter through filtering cloth in a fluted funnel. Wash with water.
9. Vacuum pump may also be used for facilitating filtration and 10 % potassium sulphate solution (hot) may be added when the filtration becomes difficult.
10. Return the residue to the digestion flask thoroughly washing the entire residue from cloth with hot water.
11. Filter into gooch crucibles prepared with thin but packed layer of ignited asbestos.
12. After washing the residue thoroughly in Gooch crucibles, with boiling water, wash with approx. 15 ml of alcohol.
13. Dry the crucibles and the contents at 110 -130° C for 2 hrs or to constant weight
14. Cool in desiccators and weigh (A)
15. Ignite the contents in a muffle furnace at 550-600° C for about 2-3 hrs until the removal of all the carbon
16. Cool and weigh (B). The loss in weight (A-B) represents crude fibre

$$\text{Crude fibre (\%)} = \frac{\text{Loss in weight recorded (A-B)}}{\text{Weight of sample (X)}} \times 100$$

Observations

Parameter	Sample No.				
	1	2	3	4	5
Wt / Vol of sample taken (X)					
Weight after all preparations (A)					
Weight after ignition in muffle furnace (B)					
Loss in weight (A-B)					
Crude fibre (%)					

For Teacher

The teacher shall tell the students

a. Method of making crude fibre extract

b. Stepwise procedure for the estimation of crude fibre

RELEVANT INFORMATION

Crude fibre is the organic residue which remains after the plant sample has been treated under standardized conditions with ether, boiling dilute sulphuric acid, boiling alkali (sodium hydroxide) and alcohol. It mainly consists of cellulose and little lignin (97%).

Principle: When the plant or food sample is washed sequentially with ether, boiling dilute sulphuric acid, boiling sodium hydroxide and alcohol, the residue left may contain crude fibre alongwith some mineral elements. This residue when ignited in muffle furnace, becomes devoid of carbonaceous material (mainly crude fibres), and is therefore estimated as loss in weight after heating to high temperatures.

Suggested Readings

Ranganna, S. 1997. Handbook of Analysis and Quality Control of Fruit and Vegetable Products. 2nd edn. Tata McGraw Hill Publishing Co., New Delhi. pp 25-26.

Sadasivam S and Manickam A 2004. Biochemical methods. 2nd edn., Interantional (P) Ltd., New Delhi pp 20-21.

❐❐❐

Exercise 29

Determination of Crude Fat

LEARNING OBJECTIVES

The students should be able to

- Understand what is crude fat
- Learn the stepwise procedure used for determination of crude fat in given samples

Materials Required

Soxhlet apparatus, petroleum ether, water bath, sample, grinder, weighing balance, filter paper, threads.

Procedure

1. Take the given sample and weigh it (A)
2. Dry the samples in an oven till there is no more weight loss and weight it (B)
3. Grind the sample into powder.

4. Place a weighed quantity of sample in filter paper, wrap the paper and tie it tightly with a thread so that no powder comes out of the paper during extraction.
5. Pour approximately 100 ml of anhydrous ether into the bottom flask. Turn the heaters on. Set the temperature at about 40 °C.
6. The ether will evaporate and circulate through the sample and again will come back to the flask.
7. Repeat the process for about 16 – 18 hrs for complete extraction.
8. Remove the sample and evaporate most of the ether from the mixture of crude fat and ether obtained during extraction.
9. Evaporate excess of ether from the crude fat sample by evaporation on a water bath. Evaporate till it is free from the fumes of ether. Record the weight (C).
10. Calculate the crude fat as per given formula.

$$\text{Crude Fat \% (fwb)} = \frac{\text{Wt. of ether soluble material (C)}}{\text{Weight of sample (A)}} \times 100$$

$$\text{Crude Fat \% (dwb)} = \frac{\text{Wt. of ether soluble material (C)}}{\text{Weight of sample (B)}} \times 100$$

Observations

Parameter	Sample No.				
	1	2	3	4	5
Initial weight of sample taken (A)					
Weight after complete drying (B)					
Weight of fat (C)					
Crude Fat (%, fwb)					
Crude Fat (%, dwb)					

For Teacher

The teacher shall tell the students

a. What is crude fat

b. Stepwise procedure for the estimation of crude fat

RELEVANT INFORMATION

Crude fat of a food represents true fat (triglycerides), phospholipids, sterols, essential oils, fat soluble pigments etc that are extractable with ether. Since the material is properly dried before loading, the water soluble materials are not extracted.

Principle: Fats are soluble in ether and are therefore extracted in it for about 10-16 hrs. The ether is evaporated form the extract and the left over of the sample is fat.

Suggested Readings

Ranganna, S. 1997. Handbook of Analysis and Quality Control of Fruit and Vegetable Products. 2nd Edn. Tata McGraw Hill Publishing Co., New Delhi. pp. 10-12.

❒❒❒

Exercise 30

Determination of Sensory Quality on 9 Point Hedonic Scale

LEARNING OBJECTIVES

The students should be able to

- Judge the minute differences in the pleasurable and un pleasurable sensory experiences
- Evaluate and rate a product on the basis of sensory quality
- Understand the conditions when 9 point hedonic rating test is used

Materials Required

Sensory scoring card, products, pen/ pencil for coding, plain water

Procedure

1. Assign codes to the samples i.e. a, b, c, …… or 1, 2, 3, ….. etc.

2. Keep the coded samples randomly (eg. c, f, a, b, d, e) in a row for evaluation
3. Start the evaluation of samples for various sensory attributes one by one.
4. Evaluator should rinse his/ her mouth after each sample so as to avoid overlapping of taste, flavour etc.
5. Assign marks from 1-9 based on pleasurable and un pleasurable experiences.
6. Compute the means for a particular attribute on the basis of the evaluation of different evaluators (replications).
7. Find out the best sample on the basis of highest average scores obtained.

Observations

HEDONIC RATING TEST

Name of Product : ....

Date : ....

Please evaluate the given coded samples for various organoleptic characteristics on 9-Point Hedonic Scale as given below

Sample Code	**1**	**2**	**3**	**4**	**5**	**6**	**7**	**8**
Skin Colour								
Flavour								
Taste								
Texture								
Juiciness								
Body								
Overall Acceptability								

Comments (if any) :..

....

Signature of Evaluator

9-Point Hedonic Scale

9 - like extremely
8 - like very much
7 - like moderately
6 - like slightly
5 - neither like nor dislike
4 - dislike slightly
3 - dislike moderately
2 - dislike very much
1 - dislike extremely

For Teacher

The teacher shall tell the students

a. about the appropriate time for doing sensory evaluation

b. the various observations and experiences, both, pleasurable and un pleasurable and assignment of scores to them.

c. the conditions under which the 9 point hedonic rating test is used.

RELEVANT INFORMATION

The 9 point hedonic rating test is used for measuring the consumer preferences of the served product (fresh or processed) on the basis of pleasurable and un pleasurable experiences. The evaluator is asked to rate the acceptability of the product on a 9 point scale from "dislike extremely" to "like extremely". This test is used when about 5-8 samples are to be evaluated for determining preferences rather than differences.

Precautions

- Select only those evaluators who are really interested and willing for sensory evaluation.
- The evaluation should be done seriously for finding preferences and each difference should be felt and experienced properly
- Avoid those evaluators, who participate in the evaluation just for consuming the product
- The evaluation should never be done immediately after taking meals or any refreshment
- The samples should be placed randomly after proper coding to avoid any biasness.
- Plain water should always be provided for rinsing mouth in between evaluation of different samples.
- Avoid evaluation of more than 7-8 samples simultaneously as the evaluator may not be able to do the correct evaluation
- The drinks should be served chilled for evaluation.

Suggested Reading

Ranganna, S. 1997. Handbook of Analysis and Quality Control of Fruit and Vegetable Products. 2nd edn. Tata McGraw Hill Publishing Co., New Delhi. pp 623-624.

❑❑❑

Exercise 31

Determination of Sensory Quality by Triangle Test

LEARNING OBJECTIVES

The students should be able to

- Understand the conditions in which this test should be used
- Conduct the triangle test for various samples
- Draw conclusions

Materials Required

Sensory scoring card, products, pen/pencil for coding, plain water.

Procedure

1. Take the given two samples
2. Make them three by duplicating/repeating either one of the two
3. Assign codes to the samples eg. AAB, ABA, ABB, BAB etc.

4. Ask the evaluators to evaluate the samples for any differences between the 3 on the given Specimen Evaluation Card.
5. Find out how many of the evaluators have chosen the odd sample correctly (say it is 16 out of 20 evaluators).
6. Now find out the degree of difference in the samples indicated by the evaluators who correctly found the odd sample (Say it is much as per 3, moderate as per 8 and slight as per 5 out of the 16 correct evaluators.).
7. Find out how many of the correct evaluators preferred A sample and how many the B sample (say 11 preferred A sample and 5 B sample).
8. After going through Annexure II you may conclude that the two samples differed significantly from each other as 16 out of 20 evaluators found the odd sample correct. But, the preference was not significant as it should have been at least 13 out of 16 judges indicating same preference so as to make it significant. So, it may be concluded that the two samples differ significantly but both are acceptable.

Observations

TRIANGLE TEST
(Specimen Evaluation Card)

Name:____________________ Date: __________

Product: ____________________

1. Here are three samples for evaluation. Two of the three samples are identical. Determine the odd samples for any difference in taste or flavour and overall acceptability. Please ignore colour.

Sample	Check odd sample
AAB	____________
ABA	____________
BAB	____________

   Did you check by guess? Yes ☐ No ☐

2. Indicate the degree of difference between like and odd samples.

   None .... Slight ....

   Moderate.. Much ....

3. Acceptability
   Odd sample preferred Yes_____ No_____
   Like samples preferred Yes_____ No_____

4. Comments, if any..

Signature of evaluator

Test Design and Results of Triangle Test

Panelist number	Sampling sequence	Difference observed	Odd sample chosen[a]	Degree of difference[a]	Preferred sample[b]
1	AAB	Yes	(A)	Much	A
2	ABA	Yes	(A)	Moderate	A
3	BAB	Yes	(A)	Moderate	A
4	ABB	Yes	(A)	Moderate	B
5	BAA	Yes	B		
6	ABA	Yes	(A)	Slight	B
7	BAB	Yes	(A)	Much	A
8	ABA	Yes	(A)	Moderate	A
9	AAB	Yes	B		
10	BAB	Yes	(A)	Slight	A
11	ABB	Yes	(A)	Much	A
12	BAB	Yes	(A)	Moderate	B
13	BBA	Yes	(A)	Moderate	A
14	BAB	Yes	(A)	Slight	A
15	ABB	Yes	B		
16	AAB	Yes	B		
17	ABA	Yes	(A)	Slight	B
18	AAB	Yes	(A)	Slight	B
19	BAB	Yes	(A)	Moderate	A
20	ABB	Yes	(A)	Moderate	A

[a] Letters in the parenthesis indicate correct identification
[b] Difference and the preference rating of panelists not correctly identifying the odd sample are not considered
A Say Drink with added natural flavour
B Say Drink with added synthetic flavour

For Teacher

The teacher shall tell the students

a. the conditions under which the triangle test is used

b. stepwise process how the triangle evaluation is conducted.

RELEVANT INFORMATION

Triangle test is generally used to evaluate the samples which differ only slightly with respect one or two of the sensory attributes and to judge the superiority or inferiority of one sample over the other. This test can not be used for evaluating more than two samples simultaneously. In triangle test 3 samples are served for evaluation. Out of these 3 samples, two are identical and one is different or odd. The panelist is asked to find out the odd sample correctly. The significance for preference and the degree of difference is evaluated at 5 or 1 % level of significance on comparison of the values with the Table (Annexure II).

Suggested Readings

Ranganna, S. 1997. Handbook of Analysis and Quality Control of Fruit and Vegetable Products. 2nd edn. Tata McGraw Hill Publishing Co., New Delhi. pp 607-609.

❑❑❑

Exercise 32

Determination of Sensory Quality by Ranking Test

LEARNING OBJECTIVES

The students should be able to

- Understand the conditions in which this test should be used
- Conduct the ranking test for various samples
- Draw conclusions

Materials Required

Sensory scoring card, products, pen/ pencil for coding, plain water.

Procedure

1. Assign codes to the samples i.e. a, b, c, …… or 1, 2, 3, ….. etc.

2. Keep the coded samples randomly (eg. c, f, a, b, d, e) in a row for evaluation
3. Start the evaluation of samples one by one.
4. Evaluator should rank the samples on the basis of intensity of a single sensory attribute according to his / her preference.
5. Compute the means on the basis of the evaluation of different evaluators (replications).
6. List the samples in increasing or decreasing order of preference.

Observations

Specimen Evaluation Card
RANKING TEST

Name : .... Date : ....

Product : ....

Please rank the following samples according to your preference for the intensity of .... (an attribute such as aroma, taste, freeness from bitterness etc,.) in the samples

Sample No.	Intensity preference
1	IV
2	III
3	V
4	II
5	I
6	VI

Comments: ....

(Signature of evaluator)

For Teacher

The teacher shall tell the students

a. The conditions under which the ranking test is used

b. Stepwise process how the ranking test is conducted.

Relevant Information

The ranking test is used when more than 3 samples are to be compared with each other with respect to the preference to a single attribute. All other attributes of the samples should remain the same.

Suggested Readings

Ranganna, S. 1997. Handbook of Analysis and Quality Control of Fruit and Vegetable Products. 2nd edn. Tata McGraw Hill Publishing Co., New Delhi. pp 609-610.

❑❑❑

Exercise 33

Visit to a Fruit Processing Unit

LEARNING OBJECTIVES

The students should be able to

- Have the practical exposure to a commercial fruit processing unit
- Understand the functioning of various equipments and machinery.
- Learn how the technology standardized in laboratory is amplified at industrial level.

Materials Required

Record book

Procedure

1. Follow the instructions given by the teacher and the factory representative.

2. Listen carefully to the explanations of the factory representative
3. Observe the functioning of various equipments
4. Ask queries wherever not understood properly.
5. Note down the observations in your record book

Observations

1. Name and address of the factory or processing unit
2. Size of the Unit (Large, Small, Cottage, Home)
3. FPO Licence No.
4. List of products manufactured
5. Height of the processing plant
6. Information regarding raw material store, finished product storage
7. List of equipments installed in the factory
8. Water supply system
9. Waste disposal system
10. Hygiene in the surroundings
11. Any other

For Teacher

The teacher shall

a. Make contacts with the processing unit manager and arrange for the visit of students
b. Tell the students not to touch anything (running machinery or products etc. unless permitted to do so).
c. Encourage the students to ask queries

RELEVANT INFORMATION

The visit should be made during the period when the unit is in functioning condition, so that the students are exposed to maximum unit operations. It should be also taken care of that the unit to which the students are being taken should not be fully mechanized because in such units the students are not exposed properly to the various unit operations followed during the manufacture of different products.

❑❑❑

References

1. Anonymous 2005. Package of Practices of Horticultural Crops. Dr Y S Parmar University of Horticulture and Forestry, Nauni, Solan HP, India 282p.

2. Fellows P 1988. Food Processing Technology- Principles and Practices. Ellis Horwood Ltd., Chichester, England. 505p.

3. FPO 1991. The Fruit Products Order 1955. Ministry of Food Processing Industries, Government of India, 49p.

4. Hulme, AC 1971. The Biochemistry of Fruits and Their Products. Vol 1 & 2, Acad. Press London.

5. Joshi VK 1998. Fruit Wines. Directorate of Extn. Edu., Dr Y S Parmar Univ. Hort. Forestry., Nauni, Solan, HP, India, 226p.

6. Kader AA, Robert RF, Mitchell FG, Reid MS, Sommer NF and Thompson JF 1985. Postharvest Technology of Horticultural Crops. Coop. Ext.Univ. Calif. DANR Berkeley 94720. 192p.

7. Kushal S, Arora JS and Bhattacharjee SK 2001. Postharvest Management of Cut Flowers – A Technical Bulletin No. 10,

AICRP on Floriculture, Div., of Floriculture and Landscaping, IARI, New Delhi 39p.

8. Lal G, Sidappa GS and Tandon GL (1998) Preservation of Fruits and Vegetables. ICAR Pub. New Delhi 488p.

9. PFA 2008. Prevention of Food Adulteration Act, 1954 with Prevention of Food Adulteration Pules 1955. Law Publishers (India) Pvt. Ltd. Allahabad, 534 p.

10. Rajput CBS and Babu RSH 1991. Citriculture. Kalyani Pub. New Delhi 368p.

11. Ranganna S 1997. Handbook of Analysis and Quality Control for Fruit and Vegetable Products. 2nd edn. Tata McGraw-Hill Pub. Co. Ltd. New Delhi 1112p.

12. Sadasivam S and Manickam A 2004. Biochemical Methods. 2nd edn., New Age International (P) Ltd. Pub., New Delhi, 256p.

13. Sharma SK, Sharma PC and Kaushal BBL 2001. Effect of storage temperature and folds of concentration on quality characteristics of Galgal (Citrus pseudolimon. Tan.) juice concentrates. *J Fd. Sci Technol* 38(6): 553-556.

14. Srivastava RP and Kumar S 2002. Fruit and Vegetable Preservation-Principles and Practices. 3rd edn. International Book Distributing Co. Lucknow 474p.

15. Ting SV and Rouseff RL 1986. Citrus Fruits and Their Products. Marcel Dekker Inc., New York, 293p.

16. Verma LR and Joshi VK 2000. Postharvest Technology of Fruits and Vegetables -Handling, Processing, Fermentation and Waste Management. Vol. I & II, Indus Pub. Co. New Delhi 1222p.

17. Wills, R.B.H., McGlasson, W.B., Graham, D., Lee T.H. and Hall E.G. (1996). Postharvest – An Introduction to the Physiology and Handling of Fruit and Vegetables. CBS Pub. & Distributors, New Delhi, 174p.

❑❑❑

Annexure - I

The Fruit Products Order 1955

The Fruit Products Order 1955 was made by the Central Government in exercise of the powers conferred by section 3 of the Essential Commodities Act 1955.

"Fruit Products" means any of the following articles

i. Synthetic beverages, syrups and sherbets
ii. Vinegar, whether brewed or synthetic
iii. Pickles
iv. Dehydrated fruits and vegetables
v. Squashes, crushes, cordials, barley water, barreled juice and ready-to-serve beverage, fruit nectar or any other beverage containing fruit juices or fruit pulp
vi. Jams, jellies, marmalades
vii. Tomato products, ketchup and sauces

viii. Preserves, candied and crystallized fruits and peels

ix. Chutneys

x. Canned and bottled fruits, juices and pulp

xi. Canned and bottled vegetables

xii. Frozen fruits and vegetables

xiii. Sweetened aerated waters with or without fruit juice or fruit pulp

xiv. Fruit cereal flakes

xv. Any other unspecified items relating to fruits and vegetables

Licensing Officer : Director (Fruit and Vegetable Preservation), Food and Nutrition Board, Department of Food, Ministry of Agriculture, Govt. of India, and includes any other officer empowered in his behalf by him with the approval of the Central Government.

Manufacturer : a person engaged in the business of manufacturing fruit products for sale and includes any person who obtained fruit products from another person and packs or labels them for sale.

After every two years the Central Government constitutes a committee to be called as *Central Fruit Products Advisory Committee* under the chairmanship of Joint Secretary to the Govt. of India, Department of Food. The Executive Director, Food and Nutrition Board, Department of Food is the Vice Chairman.

THE SECOND SCHEDULE

PART I (A)

Sanitary Requirements of a Factory Manufacturing Fruit Products

The place, where any fruit products are manufactured (hereinafter referred to as the factory), shall comply with the following requirements and in the opinion of the Licensing Officer, shall be fit for manufacturing the item or items for which the licence is granted to the manufacturer.

1. The premises shall be clean, adequately lighted and ventilated and shall be cleaned, if required, by lime washing or colour washing or painting or disinfecting or deodourising.

2. Windows, doors and other openings suited to screening shall be fly proof. The doors should have springs so that they may close automatically.

3. The equipments and the manufacturing premises approved for the manufacture of fruit products shall not be used for the manufacture of other products repugnant to the manufacture of fruit products except under the conditions given as under:

   If the licensed premises are used for the manufacture of both fruit products and fish, meat and egg products there shall be a gap of at least one month when the change is made from fish, meat and egg products to the fruit products.

   EXPLANATION

   The proposed date of changeover from production of the fish, meat and egg products to that of the fruit products, shall be intimated to the concerned Regional Officer in writing and in case of outstation factories, the intimation shall have to be given by registered letter.

4. The premises shall be located in a sanitary place and free from filthy surroundings.

5. All yards, outhouses, stores and all approaches of the premises shall be kept clean and sanitary.

6. The authorized premises shall be so constructed or maintained as to permit hygienic production and all operations in connection with preparing or packing of products shall ba carried out carefully under strict sanitary conditions laid down in the Factories Act 1934, as amended as modified from time to time. The premises shall not be used as or communicate directly with the residential premises

7. Equipment and machinery when employed shall be of such design which will permit easy cleaning. Adequate arrangements for cleaning of containers, tables, working parts or machinery etc. shall be provided.

8. No vessel, container or other equipment, the use of which is likely to ensure metallic contamination injurious to health shall be employed in the preparation, packing or storage of fruit (Copper or Brass vessels shall be always kept tinned. No iron or galvanized iron shall come in contact with fruit products).

9. The water used in the manufacture shall be potable and if required by the Licensing Officer shall be got examined chemically and bacteriologically by any recognized laboratory. The manufacturer will bear the cost of such analysis.

10. There should be efficient drainage system and there shall be adequate provisions for disposal of refuse.

11. Wherever five or more employees of either sex are employed, a sufficient number of latrines for each sex as under shall be provided.

No. of workers	No. of latrines	No. of wash basins
Upto 25	1	1
25 - 49	2	2
50 - 100	3	3
100 and above	5	5

12. Wherever cooking is done in open fire, proper arrangements shall be made for the outlet of smoke and soot.

13. No person suffering from infections or contagious diseases shall be allowed to work in the factory. Arrangements shall be made to get all the workers engaged in the manufacture of fruit products medically examined once in a year to ensure that they are free from infections, contagious and other diseases. A record of these examinations signed by a Registered Medical Practitioner shall be maintained for inspection.

    The worker engaged in the manufacture of fruit products shall be inoculated against the entric group of diseases and vaccinated against small pox once a year and certificate thereof shall be kept for inspection. In case of epidemic all workers should be inoculated or vaccinated.

14. The workers working in processing and preparation shall be provided with proper aprons and headwear which shall be clean. The management shall see that all workers are neat, clean and tidy.

PART I (B)

The factories shall be categorized as under:

Category	Installed Capacity per day	Annual Production	Min. area of manufacturing premises excluding store and office space
Large Scale	> 2 metric tonnes	> 250 metric tonnes	300 m^2
Small Scale	Upto 2 metric tonnes	> 50 and not exceeding 250 metric tonnes	
(i) Small Scale Category (A)	Not exceeding 1 tonne	50 tonnes to 100 tonnes	100 m^2
(ii) Small Scale Category (B)	Not exceeding 2 tonnes	100 tonnes to 250 tonnes	150 m^2
Cottage Scale	-	> 10 tonnes but < 50 tonnes	60 m^2
Home Scale	-	Not exceeding 10 metric tonnes	
(i) Home Scale Category (A)		Not exceeding 5 metric tonnes (Primary processing units operating in rural areas for peeling, slicing and bringing of raw mangoes for the manufacture of mango slices in brine for sale to firms licences under FPO)	10 m^2 (< 5 workers) 20 m^2 (6-10 workers)
(ii) Home Scale Category (B)		Not exceeding 10 metric tonnes (Total production of fruit products except canned vegetables)	25 m^2

WATER

Every licensee shall arrange for at least one kilo litre (1000 litre) per day of potable water and its availability shall be adequately increased as per production. Free flowing pipe water supply shall be made available in the processing hall.

Product	Min TSS %	Min % of Fruit Part	Remarks
PART II			
Fruit syrup	65	25	
Crush	55	25	
Squash	40	25	
Cordial	30	25	
Unsweetened Juice	Natural	100	
Sweetened Juice	10	85	
RTS beverages including aerated waters containing fruit juice or pulp	10	10 (5% in case of lime juice based RTS)	
Mango nectar	15	20	
Fruit Juice Concentrate	32	100	
Fruit Nectar (excluding orange and pineapple nectars)	15	20	
Orange and pineapple nectars	15	40	
Mango pulp	12		
Canned mango pulp (sweetened)	15		min. 0.3% acidity
Flavoured Sweetened Aerated Waters	8		
Sweetened Aerated Water containing fruit juice or pulp or bits	10	10	
PART III			
Barley waters (lemon, orange, grapefruit etc.)	30	25	Min. 0.25% barley starch
PART IV			
Synthetic Syrups and Sharbets	65		
Ginger cocktail, ginger beer, ginger ale	30		

PART V

Product	Variety	Special characteristics	General characteristic s
Bottled or canned fruits	Any fruit of suitable variety	Head space = Not > 1.6 cm Drained weight Not < 50% (40% for berry) Drained weight shall be determined by draining contents for 2 min., on a sieve of dimensions 20.3 x 20.3 cm having 8 mesh / 2.5 cm	Only substances that may be added are fruits, sugar, invert sugar, citric acid and water. No preservative shall be added. No artificial colouring matter shall be present, except in case of cherries and strawberries where permitted colour may be added.

contd...

Bottled or canned vegetables	Any vegetable of suitable variety	Head space = Not > 1.6 cm Drained weight Not < 55% (50% for tomato) Drained weight shall be determined by draining contents for 2 min., on a sieve of dimensions 20.3 x 20.3 cm having 8 mesh /2.5 cm	Only substances that may be added are vegetable, sugar, water oil or fat spices, sauce, citric acid and soluble calcium salts. No preservative shall be added. No artificial colouring matter shall be present, except in case of peas where permitted colour may be added.
	PART VI and VII		
Jam and fruit cheese	68	45 (25% for raspberry and strawberry)	Only substances that may be added are sugar, dextrose, invert sugar or liquid glucose, flavouring matter, ascorbic acid, citric acid, permitted colours and preservatives.
Fruit jelly and marmalade	65	45	Fruit jellies shall be made from clear fruit extracts. Marmalades shall be prepared from citrus fruit having suspended slices of peel in the finished product. Only substances that may be added are sugar, dextrose, invert sugar or liquid glucose, ascorbic acid, citric acid, permitted colours and preservatives. Should retain flavor and aroma of the original fruit.
	PART VIII and IX		
Candied Crystalized or glazed fruit and peel	70 (total sugar) reducing sugars = not less than 25% of total sugars		Only substances that may be added are sugar, dextrose, invert sugar or liquid glucose, citric acid, soluble calcium salts, flavouring matter permitted colours and preservatives.
Preserves	68	55	Only substances that may be added are sugar, dextrose, invert sugar or liquid glucose, flavouring matter, citric acid, ascorbic acid, permitted colours and preservatives.
	PART X		
Fruit Chutney	50	40	The acidity and ash contents shall not exceed 2% and 5% respectively. Dry fruit, raisins, spices, salt, onion, garlic, vinegar, acetic acid, colours and preservatives may be added.

contd...

	PART XI, XII and XIII		
Tomato juice Tomato Soup	5 7		Only substances that may be added in tomato juice are salt not in excess of 5% by weight, sugar, dextrose, malic acid, ascorbic acid, citric acid and permitted colours.. In tomato soup spices, sugar, salt, starch, butter and milk solids may be added.
Tomato Puree Tomato Paste	9 25		Only substances that may be added are common salt, citric acid, ascorbic acid, spices, permitted colours and preservatives.
Tomato Ketchup	25		Acidity min. 1.0%, Product derived only from tomatoes. Shall not contain any other fruit/ vegetable substance. Only substances that may be added are spices, salt, sugar, vinegar, acetic acid, onion, garlic and preservatives.
Sauce	15		Acidity Min. 1.2%. Product may be from tomato and or any other fruit or vegetable pulp. Only substances that may be added are fruit, vegetable pulps, juice, dried fruits, sugar, jaggery, spices, salt, vinegar, citric acid, acetic acid, malic acid, onion, garlic, flavouring matter, permitted colours other than red or any shade of red colour and preservatives.
	PART XIV, XV, XVI and XVII		
Brewed and Synthetic vinegar	Min. 3.75 g acetic acid per 100 ml	Brewed vinegar is liquid derived from alcoholic and acetous fermentation of fruits, malt, molasses, sugarcane juice etc.	Shall contain min. 1.5% w/w of total solids and 0.18 % ash. Shall not contain sulphuric acid, or any other mineral acid, lead or copper, arsenic (exceeding 1.5 ppm) and foreign substance or colouring matter except caramel.
Pickles in vinegar	Min. 2 g per 100 cc as acetic acid in fluid portion		Fluid portion of pickles shall not constitute more than 3/4th of the total content and shall not contain any ingredient other than spices, salt and sugar.
Pickles in citrus juice or brine	Min. 12 % salt. The citric acid contents in pickles in citrus juice shall not be less than 1.2%.		Only substances that may be added are spices, salt, sugar, jaggery, onion, garlic, benzoic acid and soluble calcium salts.

contd...

Oil pickle	Prepared from any fruit or vegetable of suitable variety and added any edible vegetable oil like rapeseed, mustard, olive oil etc.	Only substances that may be added are spices, salt, oils, sugar, jaggery, onion, garlic, acetic acid, turmeric, condiments and permitted preservatives.

PART XXII

Permissible Harmless Food Colours (The max. limit of any permitted coal tar colours or mixture of permitted coal tar colours which may be added to any fruit products shall not exceed 0.20 g per kilogram i.e. 0.02 %) Only ISI certified colours shall be used.

PART XXIII

Limits for Permitted Preservatives in Fruit Products

Fruit Product	Preservative	ppm (max.)
1. Fruit and Fruit Pulp or juice	SO_2	1000
2. Fruit Juice Concentrate	SO_2	1500
3. Dried Fruits (apricots, peaches, apples, pears and others) and Vegetables	SO_2	2000
Dried Fruits (raisins or sultanas)	SO_2	750
4. Squashes, crushes, fruit syrups, cordials, barley waters	SO_2	350
	Benzoic Acid	600
5. Jam, marmalade, preserve and fruit jelly	SO_2	40
	Benzoic Acid	200
6. Sweetened RTS Beverages	SO_2	70
	Benzoic Acid	120
7. Pickles and Chutney	SO_2	100
	Benzoic Acid	250
8. Tomato and Other Sauces	Benzoic Acid	750
9. Tomato puree and paste	Benzoic Acid	250
10. Syrups and Sharbets	SO_2	350
	Benzoic Acid	600
11. Crystallised, glazed, or cured fruit including peel candy	SO_2	150

Source : FPO (1991)

Annexure - II

Significance of triangle methods and paired preferences

No. of judges or judgments	Paired preferences Minimum agreeing judgments to establish significance preference (Two tailed Test) Probability level			Triangle difference Minimum correct judgments to establish significance differentiation Probability level		
	0.05	0.01	0.001	0.05	0.01	0.001
1	-	-	-	-	-	-
2	-	-	-	-	-	-
3	-	-	-	3	-	-
4	-	-	-	4	-	-
5	-	-	-	5	5	-
6	-	-	-	5	6	-
7	7	-	-	5	6	7
8	8	8	-	6	7	8
9	8	9	-	6	7	8
10	9	10	-	7	8	9
11	10	11	11	7	8	9
12	10	11	12	8	9	10
13	11	12	13	8	9	10
14	12	13	14	9	10	11
15	12	13	14	9	10	12
16	13	14	15	10	11	12
17	13	15	16	10	11	13
18	14	15	17	10	12	13
19	15	16	17	11	12	14
20	15	17	18	11	13	14
21	16	17	19	12	13	15
22	17	18	19	12	14	15

contd...

No. of judges or judgments	Paired preferences Minimum agreeing judgments to establish significance preference (Two tailed Test)			Triangle difference Minimum correct judgments to establish significance differentiation		
	Probability level			Probability level		
	0.05	0.01	0.001	0.05	0.01	0.001
23	17	19	20	13	14	16
24	18	19	21	13	14	16
25	18	20	21	13	15	17
26	19	20	22	14	15	17
27	20	21	23	14	16	18
28	20	22	23	15	16	18
29	21	22	24	15	17	19
30	21	23	25	15	17	19
31	22	24	25	16	18	19
32	23	24	26	16	18	20
33	23	25	27	17	19	20
34	24	25	27	17	19	21
35	24	26	28	18	19	21
36	25	27	29	18	20	22
37	25	27	29	18	20	22
38	26	28	30	19	21	23
39	27	28	31	19	21	23
40	27	29	31	20	22	24
41	28	30	32	20	22	24
42	28	30	32	21	22	25
43	29	31	33	21	23	25
44	29	31	34	21	23	25
45	30	32	34	22	24	26
46	31	33	35	22	24	26
47	31	33	36	23	25	27
48	32	33	36	23	25	27
49	32	34	37	23	25	28
50	33	35	37	24	26	28

Source : Ranganna (1997)

❑❑❑

Annexure - III

Temperature correction table for correcting measured TSS

Tempe-ratur'C	Quality Fraction %																	
	0	5	10	15	20	25	30	35	40	45	50	55	60	65	70	75	80	85
	Subtract from measured value																	
10	0.52	0.58	0.59	0.61	0.64	0.67	0.69	0.71	0.72	0.74	0.74	0.74	0.75	0.76	0.77	-	-	-
11	0.48	0.51	0.54	0.55	0.58	0.61	0.63	0.65	0.65	0.67	0.67	0.67	0.68	0.68	0.69	-	-	-
12	0.44	0.47	0.49	0.50	0.52	0.55	0.57	0.58	0.58	0.60	0.60	0.60	0.60	0.61	0.61	-	-	-
13	0.39	0.42	0.43	0.44	0.45	0.49	0.50	0.51	0.51	0.53	0.53	0.53	0.53	0.53	0.53	-	-	-
14	0.35	0.37	0.38	0.39	0.40	0.42	0.43	0.44	0.44	0.45	0.45	0.45	0.45	0.45	0.46	-	-	-
15	0.29	0.31	0.32	0.33	0.34	0.35	0.36	0.37	0.37	0.38	0.38	0.38	0.38	0.38	0.38	0.38	0.37	0.37
16	0.24	0.25	0.26	0.27	0.28	0.28	0.29	0.30	0.30	0.31	0.31	0.31	0.31	0.31	0.31	0.30	0.30	0.30
17	0.18	0.19	0.20	0.20	0.21	0.21	0.22	0.22	0.23	0.23	0.23	0.23	0.23	0.23	0.23	0.23	0.23	0.22
18	0.12	0.13	0.13	0.14	0.14	0.14	0.15	0.15	0.15	0.15	0.15	0.15	0.15	0.15	0.15	0.15	0.15	0.15
19	0.06	0.06	0.07	0.07	0.07	0.07	0.07	0.08	0.08	0.08	0.08	0.08	0.08	0.08	0.08	0.08	0.08	0.07
	Add to the measured value																	
21	0.06	0.07	0.07	0.07	0.07	0.07	0.06	0.08	0.08	0.08	0.08	0.08	0.08	0.08	0.06	0.08	0.08	0.07
22	0.13	0.14	0.14	0.14	0.14	0.15	0.15	0.16	0.16	0.16	0.16	0.16	0.16	0.16	0.15	0.15	0.15	0.15
23	0.20	0.21	0.21	0.22	0.22	0.23	0.23	0.23	0.23	0.24	0.24	0.24	0.24	0.23	0.23	0.23	0.23	0.22
24	0.27	0.28	0.29	0.29	0.30	0.30	0.31	0.31	0.31	0.32	0.32	0.32	0.32	0.31	0.31	0.31	0.30	0.30
25	0.34	0.35	0.36	0.37	0.38	0.38	0.39	0.39	0.40	0.40	0.40	0.40	0.40	0.39	0.39	0.39	0.38	0.37
26	0.42	0.43	0.44	0.45	0.46	0.46	0.47	0.47	0.48	0.48	0.48	0.48	0.48	0.47	0.47	0.46	0.46	0.45
27	0.50	0.51	0.52	0.53	0.54	0.55	0.55	0.56	0.56	0.56	0.56	0.56	0.56	0.55	0.55	0.54	0.53	0.52
28	0.58	0.59	0.60	0.61	0.62	0.63	0.64	0.64	0.64	0.65	0.65	0.64	0.64	0.64	0.63	0.62	0.61	0.60
29	0.66	0.67	0.68	0.69	0.70	0.71	0.72	0.73	0.73	0.73	0.73	0.73	0.72	0.72	0.71	0.70	0.69	0.68
30	0.74	0.75	0.77	0.78	0.79	0.80	0.81	0.81	0.81	0.82	0.81	0.81	0.81	0.80	0.79	0.78	0.77	0.75

Annexure - IV

List of Important Journals

Title	Publisher
A. Foreign Journals	
Acta Alimentaria	Budapest, Hungary, Central Food Research Institute
Food Chemistry	Amsterdam, Elsevier
Food Microbiology	New York Academic Press
Food Processing	Chicago, Putnam Pub. Co.
Food Research International	Amsterdam, Elsevier
Food Science and Technology Today	London Institute of Food Science and Technology
Food Technology	Chicago Institute of Food Technologists

contd...

Italian Journal of Food Science	Pinerado, Italy, Chiriotti Editori
International Journal of Food Science and Technology	Oxford, UK Blackwell Science
Journal of Food Processing and Preservation	Westport CT, Food and Nutrition Press
Journal of Food Science	Chicago IL, Institute of Food Technologists
Journal of Science of Food and Agriculture	New York, John Wiley & Sons
Trends in Food Science and Technology	Amsterdam, Elsevier
Acta Horticulturae	
B. Indian Journals	
Journal of Food Science and Technology	AFST(I), CFTRI Mysore
Indian Food Packer	All India Food Processors' Association New Delhi
Beverage and Food World	The Amalgamated Press, Mumbai
Indian Journal of Horticulture	The Horticulture Society of India, New Delhi
Indian Journal of Agricultural Sciences	Directorate of Information and Publications of Agriculture, ICAR, New Delhi
International Journal of Agriculture and Horticultural Sciences	ASHRAM, Muzaffarnagar (U.P.)
Progressive Horticulture	Hill Horticulture Development Society, Chaubattia, Uttarakhand
Journal of Scientific and Industrial Research	Directorate of Information and Publications of Agriculture, ICAR, New Delhi.

❑❑❑

Annexure - V

Important maturity indices for fruits and vegetables

Maturity Index	Fruit/Vegetable
1. Computational Indices	
a. Days from full bloom to maturity	Apple, Pears, Plum, Apricot, Peach, Almond etc.
b. Meat heat units accumulated	Apple, peas
c. Calendar Date	Almost all fruits and vegetables
d. Days from sowing/ transplantation to maturity	Pea, Carrot, radish etc.
2. Physical Indices	
a. Size, weight, external colour (visual) etc.	All fruits (citrus, mango, apple etc.) and most of the vegetables (tomato, capsicum, brinjal etc.)

contd...

b. Shape	*Angularity* of banana fingers, *Full cheeks* of mango, *Compactness* of brocolli and cauliflower curds
c. Specific gravity	Cherries, watermelon, potatoes
d. Fruit pressure/Firmness	Apple, pear, stone fruits
e. Tenderness	Peas, beans
f. Development of abscission layer	Melons, apple
g. Surface morphology	Netting in some melons, gloss/ wax in fruits like plum
h. TSS	Most fruits
i. Juice content	Citrus fruits
j. Oil content	Avocado
k. Astringency	Persimmons, dates
3. Chemical Indices	
a. Starch	Apple, pears
b. Sugar and acid	Apple, pear, stone fruits, grapes
c. Sugar : acid ratio	Pomegranate, citrus, papaya, melons, kiwifruit
4. Physiological Indices	Not used commercially, mainly used for research purpose only.

❑❑❑

Annexure - VI

Conversion of units of measurement

Primary Metric Units

Length	m
Weight	Kg
Volume	Litre

Prefixes

Pico	=	10^{-12}
Nano	=	10^{-9}
Micro	=	10^{-6}
Milli	=	10^{-3}
Centi	=	10^{-2}
Deci	=	10^{-1}
Kilo	=	10^{3}
Mega	=	10^{6}

Length

1 m	=	100 cm	=	1000 mm	=	39.37 inch
1 mm	=	1000 μ (micron)	=	10,00,000 mμ (milli microns)		
1 mμ	=	10 Angstrom units (A)				

1 inch	=	2.54 cm	=	25.4 mm		
1 foot	=	0.305 m	=	304.801 mm	=	30.48 cm
1 m	=	3.279 feet	=	1.09361 yards		
1 Km	=	1000 m	=	1093.61 yards		
1 yard	=	0.9144 m				
1 Mile	=	1.60934 Km	=	1609.34 m		
1 Km	=	0.62137 miles				
8 Km	=	5 miles (approx)				

Mass

1 grain	=	64.799 mg	=	0.0648 g		
1 ounce (oz)	=	28.3495 g	=	437.49 grains		
1 pound (lb)	=	16 oz	=	7000 grains = 453.59 g		
1 Kg	=	10 hectograms	=	100 decagrams	=	1000 g
	=	2.2046 pounds				
1 quintal	=	100 Kg				
1 metric ton	=	0.9842059 ton	=	1000 Kg		

Volume

1 ml is almost equal to 1cc (cubic centi metre) or 1 ml = 1.000028 cc

1 Litre = 1000 ml = 1000.028 cc (cm^3) = 0.001000028 m^3 = 1000028 mm^3

1 m^3 = 61,023.378 $inch^3$ = 35.3145 ft^3 = 1.3079 yd^3 (cubic yards)

1 m^3 = 999972 ml = 999.972 L = 1000 Litres (approx.)

1 ft^3 = 0.028317 m^3 = 28.317 Litres = 0.028317 Kilo Litres

1 $inch^3$ = 1.6387×10^{-5} m^3 = 16.387 ml = 16387 mm^3

British Conversions

1 Litre = 7.0392 gills = 1.7598 pints = 0.88 quart = 0.22 gallon

1 Gallon = 4 quarts = 8 pints = 32 gills = 4.54596 Litres

US Conversions

1 Litre = 8.4537 gills = 2.1134 pints = 0.1.0572 quart = 0.26418 gallon

1 Gallon = 4 quarts = 8 pints = 32 gills = 3.78533 Litres

Area

1 Ha = 10,000 m^2 = 2.5 Acres (approx.) = 12.5 Bigha (approx.)

1 Bigha = 800 m^2 (approx.)

1 Acre = 4000 m^2 (approx.)

1 m^2 = 10,000 cm^2 = 10,00,000 mm^2 = 10^{-6} Km^2

1 cm^2 = 0.155 in^2 1 in^2 = 6.451613 cm^2

1 Km^2 = 0.38610 sq. mile 1 sq. mile = 2.59 Km^2

1 m^2 = 10.75 sq. feet = 1.196 sq. yds

Temperature

For conversion of temperature recorded on degree Centigrade to degree Fahrenheit

$$^{o}F = \frac{^{o}C \times 9}{5} + 32$$

For conversion of temperature recorded on degree Fahrenheit to degree Centigrade

$$^{o}C = \frac{(^{o}F - 32)}{9} \times 5$$

Freezing Point of water	=	0° C	=	32° F
Boiling Point of water	=	100° C	=	212° F

CONVERSION CHART

——to——	Multiply by	——to——	Multiply by
cm into inch	0.3937	inch into cm	2.54
m into inch	39.37	inch into m	0.0254
m to feet	3.2808	feet into meter	0.3048
Km into mile	0.62137	mile into Km	1.60935
Kg into pounds	2.204621	Pounds into Kg	0.4536
Qunitals into Kg	100	Kg into quintals	0.01
Ounce into g	28.34954	g into ounce	0.03527

Ounce into Kg	0.02835	Kg into ounce	35.27
mm^2 into cm^2	0.01	cm^2 into mm^2	100
cm^2 into m^2	0.0001	m^2 into cm^2	10,000
cm^3 into m^3	0.000001	m^3 into cm^3	10,00,000
ml into l	0.001	l into ml	1000
ml into cm^3	1.000028	cm^3 into ml	0.999972
l into cm^3	1000.028	cm^3 into l	0.000999972
l into m^3	0.001000028	m^3 into l	999.972 (1000 approx.)
% into ppm	10000	ppm into %	0.0001

❑❑❑

Annexure - VII

Preparation of cleaning solution for glassware

Materials required

1. Potassium Dichromate
2. Concentrated Sulphuric Acid
3. Water
4. Beakers (1000 ml cap)
5. Conical Flasks (1000 ml cap)
6. Hot Plate

Method of preparation

Weigh 80 g of Potassium Dichromate and dissolve in about 300 ml water in a conical flask of 1 litre capacity. If the chemical does not dissolve properly heat a little bit over a hot plate. Cool to room temperature. Take about 460 ml of concentrated sulphuric acid in a beaker (1 Litre) and add slowly to prepared potassium dichromate solution through the walls of the flask. Label and store in a transparent reagent bottle placed near the washing sink.

Precautions

1. Prepare the solution preferably over a sink.
2. Do not add the acid at once (it will boil and splash out if done so)
3. Always add acid to water (or solution) and not water into the acid.

❑❑❑

Annexure - VIII

Preparation of chemical solutions

1. Percent solutions

Dissolve "x" g of solute in sufficient water to make the final volume 100 ml in a volumetric flask to yield "x" % solution.

Example: 2 g of Anhydrous Citric acid dissolved in sufficient quantity of distilled water (say about 60 ml) to make the final volume of 100 ml in a volumetric flask at 20° C temperature would yield 2% solution.

2. ppm solutions

Dissolve "x" mg of solute in sufficient water to make the final volume 1000 ml (1 Litre) in a volumetric flask to yield "x" ppm solution.

Example: 200 mg of citric acid dissolved in sufficient quantity of distilled water (say about 500 ml) to make the final volume of 1000 ml (1 Litre) in a volumetric flask at 20° C temperature would yield 200 ppm solution.

Note

1. Volume of the same solution would increase if the temperature is increased and would decrease if the temperature is decreased (till the freezing point). So for comparison, all solutions should be prepared at a specific temperature.

2. For conversion of ppm solution into % solution multiply by 0.0001 and for conversion of % solution into ppm multiply by 10,000.

10000 ppm	=	1%
1000 ppm	=	0.1 %
100 ppm	=	0.01 %
10 ppm	=	0.001 %
1 ppm	=	0.0001 %
0.1 ppm	=	0.00001 %

❑❑❑

Annexure - IX

Food additives and limits of their use

Food additives	**Type of foods**	**Max. permissible concentrations**
Antioxidant : BHA	*Rasogulla* and *vadas* Whole and partially skimmed milk powder Margarine RTE dry breakfast cereals	Not exceeding 0.02 (of the total fat content) 0.01% (of the finished product) 0.02 % 0.0005 % (50 ppm)
Anticaking agent: Aluminum silicate, dicalcium phosphate	Table salt, onion powder, garlic powder, soup powder	2.0 %
Sweetening agent : Saccharin	Carbonated non-alcoholic drinks	0.01 %
Sequestrant : EDTA	Canned carbonated beverages, salad dressings, margarine and sources	0.003-0.08 %

Colour	Most foods	0.02 %
Flavour : Monosodium glutamate	Meat product, soup powder	0.05 %
Sulphur dioxide or salts of sulphurous acid (Potassium metabisulphite)	Squashes, fruit pulp, crushes, jams, syrups, beer, pickles, beverages etc.	49-3000 ppm
Propionic acid and its salts	Bread and bakery	5000 ppm
Benzoic acid and its salts (Sodium Benzoate)	Chutneys, syrups, squashes, jams, RTS beverages, pickles	50-600 ppm
Sorbic acid	Beverages, cakes and icings, cheese, cider, dried fruits, margarine salad dressings, wine,	200-3000 ppm
Nisin	Cheese	1000 ppm
Nitrites	Meat products	200 ppm
Nitrates	Meat products	500 ppm

Source : PFA (2008)

❑❑❑

Annexure - X

List of suppliers of equipments and machines used in Postharvest Technology

a. **Food Processing Equipments/Machines suppliers**

1. M/s B. Sen Barry & Co., 65/11, New Rohtak Road, Karol Bagh, New Delhi 110 005, Ph: 011 – 28715553, 28713105, 28713541, 9810083440
2. M/s Bajaj Maschinen Pvt. Ltd., 7/20, 7/27, Jai Lakshmi Industrial Estate, Site IV, Sahibabad Industrial Area, Distt Ghaziabad (U.P.) Ph: 0120 – 2775119, 2775137, E mail: sales @ bajajmachines.com
3. M/s Bry-Air India Pvt. Ltd., 419-420, Udyog Vihar, Phase-III, Gurgaon, Haryana
4. M/s Burman Plant & Machinery Co. Pvt. Ltd., 36, Sarkar Lane, Kolkata700007, West Bengal
5. M/s Cantech Machines, 13, Vora Bhavan, Maheshwari Udyan, King's Circle, Matunga (C. Rly), Mumbai-400019, Maharashtra

6. M/s DSI Industries, C-I0, Devatha Plaza, Residency Road, Bangalore - 560025, Karnataka
7. M/s Eastend Engineering Co., 173/1, Gopal Lal Thakur Road, Kolkata700016, West Bengal
8. M/s Eec Cee & Co., 1, Anant Ind!. Estate, Opp. Cemet Fruit & Chemicals, Rakhial Ahmedabad-380023, Gujarat
9. M/s FMC Food Tech, 7, Shivaji Housing Society, Pune-411053, Maharashtra
10. M/s Foram Foods Pvt. Ltd., 397, Swami Vivekanand road, Vile Parle (W), Mumbai-400056, Maharashtra
11. M/s Frigoscandia Winner Food Process System Ltd., Shreesh Chambers, 3rd Floor, 25/1, Yeshwant Niwas Road, Indore-452003, M.P.
12. M/s Ganga Singh Engg. Works P. Ltd., **No.1,** Vishal Indl. Estate, Village Road, Bhandup (W) Mumbai- 400078, Maharashtra
13. M/s Gardners Corporation, 6, Doctors Lane, Near Gole Market, Post Box' 299, New Delhi- 110001
14. M/s Geeta Food Engineering, Plot C-7/1 TIC Area, Pawana M.LD.C., Thane-Belapur Road, Behind Savita Chemical Ltd., Navi, Mumbai-400705, Maharashtra
15. M/s Goma Engineering Pvt. Ltd., Majiwada, Behind Universal Pertol Pump, Thane-40060 1, Maharashtra
16. M/s Grover Enterprises, 27 Lehna Singh Market, Malka Ganj, Delhi 110 007, Ph: 011-2913714, 2923506
17. M/s Guru Nanak Engg. & Foundary Works (Regd.), 166, Indl. Focal Point, Mehta Road, Amritsar-14300 1, Punjab
18. M/s HMT Ltd., (Food Processing M/c Div.), H-2, MIDC, Chikaithana Indl. Area, P.B. No. 720, Aurangabad-431210, Maharashtra
19. M/s Jwala Engg. Co., 12, Surve Indl. Estate, Sonawala Cross Road No.1, Goregaon (E), Mumbai-400063, Maharashtra
20. M/s Macro Scientific Works, B-35/3, GT Karnal Road Industrial Area, Delhi 110 033

21. M/s Mukul Brothers Engg. Works, P.B. No. 325, Kishan Flour Mill Compound, Tirthankar Mahavir Mart (Rly. Road), Meerut-250002, U.P.
22. M/s Nirmal Services, 2254/23 Rajguru Road, Chuna Mandi, Pahar Ganj, New Delhi - 110 055, Ph: 011-23585586, 23588084, 9891193434, E mail: virendrakapoor@ hotmail.com
23. M/s Orbit International Technologies (P) Ltd., 404, Taramandal Complex, Secretariat Road, Hyderabad -500004, A.P.
24. M/s Ratan Equipments 69, Lake View Road, Kamakati St., West Mambalam, Chennai-600003, Tamil Nadu
25. M/s Rita Bottling Machines Ltd., 1B & 1C, Suvarna Darshan, 47, 2nd Main Road, Gandhi Nagar, Adyarn, Chennai-600020, Tamil Nadu
26. M/s Sandeep Instruments and Chemicals, 3229, Ranjit Nagar, New Delhi 110 008, Ph: 011-65450982, 25843541, E mail: sanco_lab@hotmail.com, sanco_lab1@sify.com
27. M/s Sangram Engg. Ltd., B-5, Super Con, Opp. LT.L, Aundh, Pune411007, Maharashtra
28. M/s Sri Rajalakshmi Industrial Agency, 57(30/1), Silver Jubilee Park Road, Rajalakshmi Corner House, Post Box No. 6690, Bangalore- 560002, Karnataka
29. M/s Veenu Hitech, Plant Manufacturing Pvt. Ltd., F-6 St. Soldier Tower, G Block, Vikas Puri, New Delhi-ll0018
30. M/s Wintech Taparia Ltd., 25/1, Yeshwant Niwas Road, Shreesh Chambers, III Floor, Indore-452003, M.P.

b. Oil Extraction and Processing Equipments / Machines suppliers

1. M/s Sardar Engineering Co., 77/148 Latouche Road, Near Fire Brigade, Kanpur 208 001 Uttar Pradesh. Ph: 2367333, 2520579
2. Kisan Krishi Yantra Udyog, 18-C, Pokharpur, Lal Banglow, Kanpur - 208 010, Ph: 2450753, 2450819, E mail: kkyu_knp@rediffmail.com

3. Sardar Agencies, 87/8, Kanpur Tannery Compound, Bhanna Purva, Kanpur.

c. **Wrapping/Filling/Packaging Equipments / Machines suppliers**

1. M/s Aarkay Wrapping Machines Pvt. Ltd., **1,** Hormurz, 131, August Kranti Marg, Mumbai-400036, Maharashtra
2. M/s APT Packaging, 184/5007, Pantnagar Ghatkopar (E), Mumbai400075, Maharashtra
3. M/s B. Sen Barry & Co., 65/11, New Rohtak Road, Karol Bagh, New Delhi 110 005, Ph: 011 - 28715553, 28713105, 28713541, 9810083440
4. M/s Bajaj Process Pack Maschinen Pvt. Ltd., 7/20, 7/27, Jai Lakshmi Industrial Estate, Site IV, Sahibabad Industrial Area, Distt Ghaziabad (U.P.) Ph: 0120 - 2775119, 2775137, E mail: sales @ bajajmachines.com
5. M/s Bombay Engineering Industry, R. No.6 (Estns.), Sevanthibhai Bhavan, Chimatpada, Marol Naka, Andheri (E), Mumbai-400059, Maharashtra
6. M/s Europack Machines (I) Pvt. Ltd., Akash Business Centre, CST Road, Kurla 0N), Mumbai-400070, Maharashtra
7. M/s Multipack Systems Pvt. Ltd., 2nd Floor, Patrict Complex, EBora Park, Vadodara-390007, Gujarat
8. M/s Packaging Machines of India, 8-3-229/5/3/7, Jai Bharani Nagar, Yusufguda Check Post, Hyderabad, A.P.
9. M/s Reliance Packaging, 155-A, Ekta Enclave, Peeragarhi, Main Rohtak Road, New Delhi - 110041
10. M/s Sandeep Instruments and Chemicals, 3229, Ranjit Nagar, New Delhi 110 008, Ph: 011-65450982, 25843541, E mail: sanco_lab@hotmail.com, sanco_lab1@sify.com.

d. **Refrigeration and Cold Storage Equipments / Machines suppliers**

1. Acmas Technocracy Pvt., Ltd. 312-313, Vardhman Capital Mall, L. S. C. Gulabi Bagh, Delhi - 52, Ph: 9312219738, 011-22942133, E mail: info@acmasindia.com

2. M/s Anand Refrigeration Co. Pvt. Ltd., D-Lajpat Nagar-I, New Delhi-II 0024
3. M/s Blue Star Ltd., Kasturi Bldg., Mahan T. Advani Chowk, Jamshedji Tata Road, Mumbai-400020, Maharashtra
4. M/s Carrier Refrigeration (P) Ltd., 700/701 A, Lado Sarai, Aurobindo Marg, Mehrauli, New Delhi-II 0030
5. M/s Greenfield Agencies Shop No.6, Anuradha, Opp. Golf Club, Rest House, Tidke Colony, Nasik-422002, Maharashtra
6. M/s Industrial Refrigeration Pvt. Ltd., 901, Maker Chambers-V, Nariman Point, Mumbai-400021, Maharashtra
7. M/s Kooling System, 35, Mannar Reddy Street, T. Nagar, Chennai- 600017, Tamil Nadu
8. M/s Western Refrigeration Ltd., 4, Ready Money Terrace, Dr. A.B. Road, Worli, Mumbai-400018, Maharashtra
9. SV Instruments Analytical Pvt., Ltd., 3269 Ranjit Nagar, Near Pusa Gate, New Delhi 110 008, Ph: 0141-25843571, 25843572.

❑❑❑

Annexure - XI

Conduct of a student / researcher in a laboratory

Laboratory is a place where a student/researcher undertake his/her research for various projects and degree programmes. The success or failure of any research experiment depends largely on the conduct of a student in the laboratory. Here are some of the points one should remember while working in lab.

- Do not disturb any equipment, chemical etc. in the lab unless you are working with it.
- If you do not know how to handle any equipment, ask the teacher or Lab. Incharge
- If any undesirable happening occurs in the lab. like breakage of glassware, problem with some equipment, injury due to improper handling of chemicals, inform the Lab. Incharge or the Lab. staff immediately. Do not try to hide such incidents.

- Do not shift the equipments from one place to other in the lab.
- Perform all tests or experiments at the particular place mean for that purpose.
- Never leave your data record book un attended in the lab as somebody else may misuse it.
- Always check for the availability of chemicals and glassware etc. well in advance to the beginning of your actual experiments. If any thing is not available in the lab. give your requirement to the concerned staff so that the same could be procured in time.
- Do not disturb the experiments of other students in the lab.
- Do not keep any chemical un labelled.
- In order to get reliable results, the glass ware being used in the lab should be properly washed and dried. Always wash your glassware on your own. Do not trust on the glass ware washed by others.
- Wash the glassware immediately after finishing the experiment by using the following solution. Dry them in oven at 70 $\pm$ 2°C temperature.
- Always keep a Max. / Min. temperature thermometer and hygrometer in the laboratory and record the observations daily especially during the storage experiments.
- While weighing chemicals over a balance, some quantity of chemical fall over the balance. Immediately clean it with the help of a hair brush so that it does not react with the surface or with other chemicals falling bovver it.
- Always use distilled water for preparation of standard solutions used in various analysis. Never make chemical solutions in plain tap water.

- Always do the weighing and pippetting carefully as any error committed during this may make your analysis erroneous.
- Switch off all the power points which are not in use while you leave the lab.
- Close the windows and doors properly before leaving the lab.
- Get your results checked regularly so that any mistake, if happening, may be corrected at the earliest.
- If you wish to work in the lab during off hours, take permission of the Lab. Incharge.
- Do not keep the chemicals open. Keep them exactly where they are supposed to be.

❑❑❑

Index

B

C

D

E

F

G

N

O

P

Q

R

S

U

V

W

Y

Z